**1 Book + 1 Disc**

PATENT #680810     JM COMPANY     US PATENT #4450731

*To Barbara, with love*

# Starry Night

## How to Skywatch in the 21st Century

Written by John Mosley

Foreword by David Levy

An Original Publication of ibooks, inc.

Edited by Mike Parkes
Cover photo composite © 2000 Dennis Di Cicco and Terence Dickinson

Space.com
284 Richmond St. E.
Toronto, ON
M5A 1P4, Canada
(416) 410-0259
www.space.com
www.starrynight.com
support@starrynight.com

ibooks, inc.
24 West 25th Street
New York, NY 10010
www.ibooksinc.com

ISBN 0 –7434-2395-X

First Printing October 2001

10 9 8 7 6 5 4 3 2 1

Printed in the U.S.A.

# CONTENTS

# FOREWORD

## Friends of the Night

*A somewhat embarrassing thing happened to me on a clear evening about ten years ago. I was outside under the stars, hunting for comets through my 16-inch telescope, essentially minding my own business. Steve and Donna O'Meara, two close friends, were visiting at the time. As Donna explains, when she came outside she heard me saying, "Hello, how are you?" and then a few minutes later, "haven't seen you in a while." Donna thought I was talking over the observatory telephone, but as she approached the observatory, it was obvious to her that I wasn't. I was actually talking to my celestial friends, objects in the sky I was encountering in my nightly search for comets.*

# Jupiter

Not known for talking to myself, I was a little unnerved when Donna entered the observatory. But I really do see the objects in the night sky as personal friends. After all, I've looked at some of them thousands of times! Take Jupiter, for instance. This big planet was the first thing I saw through a telescope, a childhood memory from the summer of 1960 that stands out in my mind. That story began when my dad and my uncle walked into our living room with a long box. Inside were three cast-iron tripod legs, a platform, and a long tube. As they went about putting the pieces together, I looked on amazed at this surprise gift of my first telescope. "We were going to wait till your bar mitzvah," Uncle Sidney said about the event that was still an eternal nine months off, but "your parents told us how anxious you are to have a telescope."

On the night of September 1, 1960, my parents and I were outside with my new telescope, which I dubbed Echo. We had no idea what star to look at first. We had no guide – no star chart, no magazine, and certainly no software to show us where to look. Thus, I decided to point Echo toward the brightest star in the sky. As I focused the image, it grew sharper, and I could tell that what started as a round, flat blob of light was shrinking not to a star at all but a beautiful sunlit world with bands of dark clouds and three nearby stars that could only be moons of the planet Jupiter. My parents and I stood there, utterly fascinated, as we gazed on a world half a billion miles away.

A unique world, Jupiter proved its friendship to me over and over. From January 21 to February 24, 1963, while far from home as a patient at Denver's Jewish National Home for Asthmatic Children, I spent clear-sky evenings painstakingly recording the changing positions of Jupiter's big moons – Io, Europa, Ganymede, and Callisto – as they completed their orbital dances about the planet. From Io's quick two-day run around Jupiter to Callisto's stately multi-week loop, I was able to figure out which moon was which just from the chart I made. After that, Jupiter became the traditional first object I would look at through each new and bigger telescope. As I watched Jupiter's moons in 1963, I could not possibly be aware that a tiny, secret moon was also busily circling the planet in wide, two-year-long orbits.

## A Friend Pays Back

On March 23, 1993, I was far from Montreal or Denver. High atop Palomar Mountain in California, Gene and Carolyn Shoemaker and I were waiting out a cloudy night while trying to conduct our search for comets and asteroids. With some clearing, we decided to resume our photography with a field that happened to be close to Jupiter – so close that Jupiter's glare was making it hard for me to see the star on which I was to guide the eight-minute-long time exposure. In order to use photography to search for comets, we needed to have two photographs, taken about 45 minutes apart, of each area of sky we search. But 45 minutes after I took the Jupiter picture, the sky was cloudy again. We had to wait an extra hour before a small hole in the clouds allowed us to sneak in the second Jupiter picture.

A day and a half later, Carolyn Shoemaker was using a stereomicroscope to scan the two pictures. Near the center of the two pictures she found what she described as a squashed comet – instead of a coma and tail, there was a bar of cometary light, with several tails pointing northwards. Called Shoemaker-Levy 9, this comet had grazed by Jupiter just a few months before discovery. Just as the Moon pulls on the waters of Earth, Jupiter's gravity stretched this comet so strongly that it fell apart into many fragments, a "string of pearls," each with its own tail.

The comet's strange appearance was only the herald of what was to come. Instead of orbiting the Sun, as every other known comet has done, Shoemaker-Levy 9 was tracing a path around Jupiter, and had been since about 1929, as the secret moon no one knew about until we discovered it in 1993. On May 22, 1993, the International Astronomical Union issued a major announcement: Comet Shoemaker-Levy 9 was in its final orbit, and would collide with Jupiter during July of 1994. That gave the astronomical community fourteen months to plan the biggest observational campaign ever put together in the history of astronomy for a single event. By the night of the first impact on July 16, 1994, virtually every major telescope on Earth was pointed toward Jupiter. In space, the Galileo spacecraft en route to Jupiter, and the Hubble Space Telescope, watched as a fragment of Shoemaker-Levy 9, traveling at 134,000 miles per hour, tore into Jupiter's atmosphere. By the end of crash week, Jupiter's southern hemisphere was blackened by dark clouds that lasted for months. I still look at Jupiter, now a most special friend, every chance I get.

## Of Comets and Shooting Stars

To be a part of Gene and Carolyn Shoemaker's search for comets was a privilege. Comet Shoemaker-Levy 9 was one of 13 comets we found together. Comets are not among the objects that are normally plotted on a star atlas or computer program, because they move across the sky and never stay in the same place. As Leslie Peltier, an Ohio farmboy who discovered a dozen comets, wrote years earlier in his autobiography *Starlight Nights*, comets can teach us about our world in a special way:

Time has not lessened the age-old allure of the comets. In some ways their mystery has only deepened with the years. At each return a comet brings with it the questions which were asked when it was here before, and as it rounds the Sun and backs away toward the long, slow night of its aphelion, it leaves behind with us those questions, still unanswered.

To hunt a speck of moving haze may seem a strange pursuit, but even though we fail the search is still rewarding, for in no better way can we come face to face, night after night, with such a wealth of riches as old Croesus never dreamed of.

Comets are not the only objects in the solar system that move across the sky in deliberate ways. In their wake are countless specks of dust that travel through the solar system and often plow into Earth's atmosphere. As they strike, they cause the rarefied gases of the upper atmosphere to heat to the point of leaving an incandescent glow. It was one of these glows that I saw, after a Fourth of July celebration in Vermont in the late 1950s. It was my first meteor, and it ignited a lifelong interest in watching meteors.

A few years later I was out watching meteors again. August 12, 1962: The sky above Quebec's Jarnac Pond was inky, the night silent except for the occasional croak of a frog and the hoot of an owl. I could see more stars than I could count, and the Milky Way arched right overhead. I was fourteen, and for the past two years I'd been increasingly drawn to astronomy. "You're going to sit on the dock all night long? *Until dawn?*" My astonished grandparents couldn't believe my plan. On that lazy afternoon of August 12, we were watching clouds pass by and talking about the very bright meteor we had seen the previous evening. I explained to my grandparents that meteors in a shower like the Perseids appear to come from a single point in the sky, called a radiant. I added that this is due to an effect of perspective, just like looking down a railroad and watching the tracks converge. But then came the punch line: as Perseus rose higher, we would see more meteors; the shower would be strongest just before dawn. That meant, I insisted, that I was going to stay up all night. That evening I began my vigil under a partly cloudy sky, but as the night went on the clouds gradually dissipated. With each passing hour the numbers of meteors increased. One of them appeared to split in two. Others left beautiful sparkling trails. Alone on the lake that night, I was treated to a personal show of fireworks in the sky. When dawn finally came I had logged 112 meteors.

## Star Light, Star Bright

Moving outward from our solar neighborhood through space and time, we come to the "fixed" stars – points of light that are the foundation of the night sky, and the lynchpins of any star map program. Although the first stars I ever identified were those of the Big Dipper, it was the summer triangle of Vega, Deneb, and Altair that really caught my attention in those early years. But after a few years another kind of star caught my attention, a star that didn't just shine there, but which actually put on a performance. Variable stars, as they are known, are stars that change in brightness. Amateur astronomers can monitor these changes in a star by comparing, from night to night or week to week, its brightness with that of nearby stars that do not vary. In this way we become familiar not just with where a star is in the sky, but with how it behaves. Since 1911 the American Association of Variable Star Observers has been collecting amateur observations of these stars.

Of all the variable stars I have seen, my favorite is TV Corvi. When I started observing variables in 1964 I hadn't heard of this star in Corvus the crow, but neither had anyone else, save Clyde Tombaugh, the discoverer of Pluto. As part of his search for new worlds he had recorded, in 1931, the sudden appearance of this exploding star. While doing research for Tombaugh's biography, I found his records of this exploding star on a plate where no star was seen on the two earlier plates. Although Clyde reported the star's outburst to his superior at Lowell Observatory, it was never followed up. I decided to investigate. At the Harvard College Observatory is one of the world's best archives of photographic sky surveys. Although they had a plate taken of the Corvus region a few days after Clyde's exposure, the nova was absent from it. Discouraged, I looked at some other pictures of Corvus taken over many years. On the tenth one, I found Tombaugh's star again in outburst!

I decided next to examine every one of the 360 Corvus plates in the Harvard archives. Over the next three days I spotted ten more instances of the star exploding. Shortly after I returned home, I began observing the star through my own telescope. On March 23, 1990, I looked through the eyepiece and saw Tombaugh's star for myself for the first time. Fifty-nine years to the day after its first recorded outburst, this star's light was for the first time being directly witnessed by human eyes. TV Corvi, as the star was later named, is a pair of stars orbiting each other. The smaller one is a white dwarf probably no larger than the Earth. Over several months it steals hydrogen from its neighbor star. The hydrogen collects in an "accretion disk" and every few months it ignites. The neighbor star is probably only a few times bigger than Jupiter. If it is that small, it is an example of a kind of body that is at the dividing line between a star and a planet. Too big to be a planet but too small to undergo nuclear fusion, the other object is a brown dwarf, a star "wannabe."

The two suns orbit each other in an area of space smaller than our own Sun.

The sky is full of "cataclysmic variables" like TV Corvi, stars that erupt every few months. The explosion actually does no damage to the star, which may   continue to blow off hydrogen in this way thousands of times. One of the most popular is SS Cygni, a binary system whose outbursts have been followed by the AAVSO's amateur observers since 1896. Although the eruptions occur every two months on the average, you never know when this type of variable star will next explode. You just need to visit it every clear night.

## Above and Beyond

My earliest years of stargazing also included the remote galaxies. On September 9, 1962, I decided to search for the Andromeda Galaxy, a maelstrom of hundreds of billions of suns. Following the directions provided by a star chart, I sighted a bright star with two fainter stars that pointed the right direction. I gently moved the telescope a field over, then another, and soon spotted a faint patch of hazy light. From my own home, in the middle of light-filled Montreal, I was looking out into the Universe at a galaxy two million light years away. Many times since then I have looked on this old friend. Through a much larger, 16-inch telescope, and under the much darker sky at my home in Arizona, I now see the Andromeda Galaxy as a magnificent spiral-shaped maelstrom of stars, cloudy hydrogen-rich regions, and two smaller companions. But none of these modern views can compare with the feeling of discovery I felt the night I was transported through space to what I thought was the very edge of the Universe.

# Van Gogh's Starry Nights

We all see the Universe in our own ways. I saw a distant galaxy as a patch of hazy light; Van Gogh might have seen a distant galaxy in a very different way, and painted it in one of his most famous works of art. Two of Van Gogh's paintings have "Starry Night" in their titles. The earlier one, completed in the autumn of 1888, was *Starry Night over the River Rhône*. It shows the seven stars of the Big Dipper seemingly gathering water from the river. The other *Starry Night* is one of the most famous works of art in existence. I have heard that the stars that he depicts variously refer to Venus and the Moon, which were in the morning sky at the end of June 1889, the month he painted it. More intriguing are the swirls of hazy starlight that cross the sky over the town. On a first look, these swirls, and the stars that they surround and cross, seem to be ordinary stars seen through the eyes of someone intoxicated or extremely nearsighted. The artist, however, might have had much more in mind. To understand this, we need to travel in space and in time from St. Rémy, where the artist lived and saw the stars he painted, to 1850 and the town of Birr, Ireland, where amateur astronomer Lord Rosse was observing through the world's largest telescope, a reflector with a mirror seven feet wide.

With the mirror end on the ground, the mighty telescope's long bulky tube was slung with ropes between two large brick walls. Reaching his eyepiece by means of a staircase, Rosse observed and drew all kinds of celestial objects including Messier 51, the Whirlpool, located just south of the Big Dipper's handle. He drew the swirls of light from the Whirlpool. It is quite possible that Van Gogh saw Rosse's published drawing, for the swirls of light in his painting bear an uncanny resemblance to the Whirlpool.

Now known to be a vast spiral galaxy some fifteen million

light years away, M51 is 50,000 light years across, and shines with the intensity of 10 billion suns. It can be spotted as a diffuse patch of light through small telescopes, but through large telescopes the vastness of the spiral structure is spectacular. Some years ago Wendee, my wife, and I explored the details of M51 with the 61-inch Kuiper telescope in the Catalina north of Tucson. The galaxy seemed to have no end as it spiraled outward from its center. It was a staggering sight, unchanged from the beauty that Rosse had seen so long ago. It was the first time Wendee actually saw the separate arms unfolding from the galaxy's center, and we were both entranced.

## A Friend Remembers a Friend

As I write these words, I have just returned from a trip to Zambia, where I saw a total eclipse of the Sun, Mars overhead at one of its closest approaches to Earth, and an area of the sky I can never see from home. It's the area of our own Milky Way that begins with Alpha Centauri, the nearest star to the Sun, and continues through the dark Coal Sack Nebula, through the Southern Cross, and on to the Eta Carinae region of the Milky Way. I wish I could cut out that section of sky, and paste it against the sky I see from southern Arizona. "If the sky has a soul," my friend Steve O'Meara said, "it lies in that part of the sky."

At the western edge of that region is Eta Carinae. Seen visually through a telescope, Eta Carinae is simply outstanding, at low power due to the complexity of structure and color in the nebula that surrounds the star and at high power because of the gaseous structure around Eta itself. These are reasons enough to look at Eta Carinae at every opportunity, but the reason I love it so much is a personal one. Eta Carinae reminds me of Bart Bok, the Milky Way specialist

I knew and respected.

When Bart Bok arrived at Harvard in 1930, he was working on his Ph. D. dissertation about Eta Carinae, a system he thought to be quite unique in the galaxy. During this time he married astronomer Priscilla Fairfield, and over the course of their lives they spent some time in the Southern Hemisphere observing, photographing, and otherwise studying this beautiful area. As Priscilla grew older, she began to suffer from a failing memory, and Bart left his own astronomical career to care for her. On November 15, 1975, Priscilla and Bart attended the opening of the Flandrau Planetarium in Tucson. They arrived early, so that Bart could show Priscilla the planetarium's display of photographs of various parts of their beloved Milky Way. As the couple walked past the pictures, Priscilla stopped at the one showing Eta Carinae. "You know, Bart," Priscilla said, "when I am gone, that is where I am going. I will ask St. Peter to give me a front row seat right at the center of the nebula. I'll see stars forming right before my eyes!"

Overwhelmed by this profound merging of his two greatest loves, Bart tried to hold back tears as he hugged Priscilla. "Eta Carinae," Priscilla repeated as they walked into the planetarium theater for the opening ceremony, "that is where I want to be."

Priscilla died just four days later. In the years after that Bart often told this story as he recalled his own lifelong love affair with the night sky. For me, Eta Carinae is a powerful connection with a deeply touching human story and its relation to one of the most beautiful places in the night sky. Bart died in 1983, and I know he is out there, in the Eta Carinae nebula with Priscilla. When I looked at the Eta Carinae nebula last month, I thought of this wonderful friend who managed to forge such a personal relationship with the sky.

For me, the sky is all about friendship, both of the planets, stars, and galaxies that I see so often, and of the people I have met, or read about, during this association. It is about placing your heart and soul on canvas as you draw the swirls of a distant galaxy into the sky over a small French town. As you read *Starry Night*, may you use its pages to forge your own relationship with the sky. May it be one that lasts a lifetime and becomes full of cosmic friends.

—David Levy
Vail, Arizona
July, 2001

# INTRODUCTION

The sky is enormous, distant, and filled with mysterious things. It doesn't resemble anything we encounter in our daily lives. There is movement and change in the sky, but generally at a rate too slow for us to notice. To many, the sky is intimidating. It may seem inaccessible. But it is superbly fascinating. You bought this book and software package because you would like to get to know it better. Welcome – the adventure is about to begin.

Two hard facts about the sky are that (1) changes in it happen slowly, and (2) we cannot cause it to change. We cannot turn it and look at it from a different angle; we cannot speed up the rotation of the Earth or the motion of the planets through the constellations; we cannot cause an eclipse to happen; we cannot view the sky from another place on Earth (unless we actually go there); we cannot see what the sky looked like long ago or in the distant future. We are stuck in the here and now.

That is the value of this book and software package. Starry Night lets you manipulate the sky. You can move backward and forward through time as far and as fast as you wish; you can move to any spot on the Earth and to any planet; you can see imaginary lines in the sky (such as the paths of the planets and constellation boundaries); you can discover future events and watch them before they actually happen, or replay famous events from the past. You can take control of the sky. As you go, you will gain a far better understanding of how the sky works than you can as a passive observer.

Then, when you actually step outside on a clear night, you will have a greater understanding of what you see. It will be fun. Starry Night lets you stargaze on cloudy nights and even during the daytime, but there is no substitute for stepping outside at night and seeing the real thing. Starry Night will help you appreciate what you see when you are outdoors on a starry night, but don't forget to actually go outside and look up!

# How to Use This Book

Before using this book, we recommend that you install Starry Night from the CD, following the instructions on the CD itself. It is also recommended that you play around with Starry Night and become familiar with how it works. You can access an electronic manual by clicking the Start button on your desktop and choosing "Programs | Starry Night Starter | Manual" (Windows) or by double-clicking the file "Manual.pdf" in your Starry Night Starter folder (Macintosh). Once you know the basics of using Starry Night, you are ready to use this book.

*Starry Night, How to Skywatch in the 21st Century* is organized into five sections. Starry Night example files will be used in almost every section to complement the material in the text. These files can be opened by clicking the "Go" menu in Starry Night and then choosing "Companion Book." The files are subdivided by chapter. Just click on the file you are interested in to launch this file.

The first section covers the essentials of understanding the night sky. Several step-by-step exercises let you get familiar with Starry Night and show how it can help you learn about astronomy. Newcomers to the hobby may want to read Section 1 and then spend some time using Starry Night and making their own observations, before returning to the rest of this book.

The next four sections go into more detail.

Section two will help you get the most out of your observing, no matter what you are looking at. It has observational tips, hints on choosing binoculars and telescopes, and a good description of the various astronomical coordinate systems.

Section three covers the three different motions of the Earth which account for much of the variation we see in the sky.

Section four looks at other objects in our solar system, including our Moon and the nine planets.

Finally, the fifth section takes you into the deep, describing stars, galaxies, and more.

# Section 1

# Astronomy Basics

THIS SECTION of *Starry Night* will give you a quick overview of the basics of astronomy. Chapter 1.1 has general information, while Chapter 1.2 looks at the motion of the Earth and Chapter 1.3 introduces the other bodies in our solar system.

# THE NIGHT SKY

## The Constellations

Stepping outdoors at night in a dark place far from the bright lights of the city, we are amazed at the number of stars we can see. Bright stars and dimmer ones, tightly knit clusters with many points of light and dark patches with nothing at all. Stars, stars everywhere, stretching off to the horizon in all directions, as far as the eye can see. But which one of those bright points is the North Star? Is that reddish dot Mars or a relatively cool star? And isn't the International Space Station up there somewhere tonight? Wouldn't it be nice to have a map?

In fact, the sky has been mapped. Astronomers divide the sky into 88 different non-overlapping areas called *constellations*. You can think of the constellations as the "countries" on the surface of the sky. Each constellation is an area of the sky. Just like countries

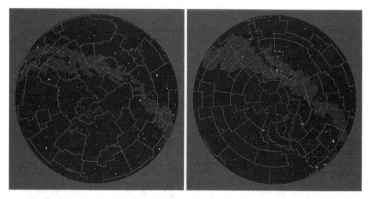

Constellation maps of the northern (left) and southern celestial hemispheres.

on Earth, some constellations are bigger than others. Everything inside the boundaries of a constellation is considered to be part of that constellation, regardless of its distance from Earth. A constellation is actually a wedge of the universe, with Earth at its center, that extends to the farthest reaches of space.

The brighter stars in each constellation make a unique pattern, and we look for this pattern to identify the constellation. To help make this identification easier, people through the ages have "connected the dots" of stars to draw recognizable patterns called constellations, which they then named. Sometimes the pattern actually looks like the object after which it is named; more often it does not. One point of confusion among newcomers is that the boundaries and the connect-the-dot stick figures share the same constellation name. If you read that the Sun is "in" Scorpius, the

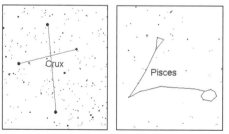

The constellation Crux has the appearance of a cross, but Pisces bears little resemblance to a fish.

scorpion, it means that the Sun is inside the official boundaries of the constellation Scorpius, but it is not necessarily inside the stick figure that outlines a scorpion. Starry Night can show you the constellation boundaries, stick figures, or both.

Mars, Jupiter, and Saturn are all considered to be in Aries, though they are outside its stick figure.

The constellations are the fundamental units of the visual sky, yet they are imaginary. They were **invented**. People created them long ago – in some cases in prehistoric times – and often for reasons that we might think strange today.

One constellation leads to another, so to speak. Use easily-recognizable star patterns to find more obscure ones. The two stars at the end of the bowl of the Big Dipper, for example, point to the North Star and to the Little Dipper, while the five stars of the handle of the Big Dipper point to the star Arcturus in Bootes.

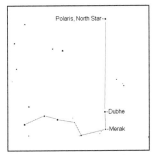

The Big Dipper's Merak and Dubhe make a straight line leading to Polaris.

The three belt stars of Orion point west to Aldebaran in Taurus and east to Sirius in Canis Major. It is easy to draw lines and arcs from star to star and "star hop" from constellations you know to those you are still learning.

As in so many endeavors, don't bite off too much. Don't try to learn the entire sky on your first night out.

The belt of Orion is sand-wiched between the bright stars Sirius and Aldebaran.

Begin by learning the major constellations and then fill in the small and obscure ones as the need and challenge arises.

A curious feature of the constellations is that once you know them, they are hard to forget. Like riding a bicycle, once you know them you tend not to forget them, and they will remain with you for a lifetime. If you know the constellations, the sky is never an unfamiliar place and you will be at home under it wherever you go. The constellations are described in much more detail in Section 5 of this book.

# Coordinate Systems

Although the constellation boundaries are a good starting point for describing the approximate location of an object in the sky, astronomers need a more precise method of specifying an object's position. There are several different coordinate systems in common use, and the most intuitive is the ***horizon coordinate system***, also known as the Alt-Az system.

"Alt" is short for ***altitude***, which is an object's height above the horizon in degrees (0 to 90). An object with an altitude of 0° is right along the horizon, while one with an altitude of 90° is directly overhead in the sky. The imaginary point in the sky directly overhead is known as the ***zenith***. Knowing the altitudes of objects at different times will help you plan your observing sessions. The best time to observe something is when it has an altitude between 20° and 60°. This is because objects (especially planets) near the horizon can be blurred by intervening air currents, while objects with an altitude greater than 60° are tough to look at without straining your neck!

**Avoid observing planets near the horizon.**

"Az" stands for *azimuth*, an Arabic word (as is altitude) for the position along the horizon; "bearing" is an alternative navigational term. Azimuth is measured in degrees from north, which has an azimuth of 0°, through east (which has an azimuth or a bearing of 90°), through south (180°) and west (270°). The *local meridian* is the imaginary line which divides the sky into an eastern and a western half. It extends from the southern point on the horizon through the zenith to the northern point on the horizon. The half of the meridian line that is south of the zenith has an azimuth of 180°, while the half north of the zenith has an azimuth of 0°.

Once you have the altitude and azimuth of an object, you know where to find it in the sky. A compass is helpful in determining the azimuth of your viewing direction (compasses point to "magnetic north," which is not exactly the same as "true north," but the two directions are practically identical unless you are observing from a far northern latitude). Once you have your bearing, it is just a matter of looking up to the proper altitude. With practice, you will quickly learn how high above the horizon is 10°, how high is 30°, and so on. Starry Night can show you the altitude and azimuth of any object at any time (just double-click on the object to bring up its Info Window with this information) and can also mark the zenith and **nadir** points and/or the meridian line (choose the appropriate option from the "Guides" menu).

Although the horizon coordinate system is the easiest to understand, it is not necessarily the most useful system. This is because it is a "local" system, and the coordinates depend on your personal location. Jupiter, for example, will have one set of horizon coordinates for an observer in California, a second set of coordinates for an observer in Texas, and a third for an observer somewhere else. To make things worse, the coordinates change constantly over time as the sky rotates. For these reasons, stars and other objects in the sky are more commonly identified using the *equatorial coordinate system.* This system gives one unique set of coordinates for an object, which is the same anywhere on Earth. The equatorial system is described in Chapter 2.2.

# Angles in the Sky

Astronomy is full of angular measures, and it is very handy to have a conceptual feeling for what these angles mean. A circle is divided into 360°, so the distance from the horizon to the zenith is one quarter of a circle or 90°. It is also 90° along the horizon from one compass point to the next (east to south, for example). Familiar constellations will help you visualize smaller angles. The distance from the end star in the handle of the Big Dipper to the end star in the bowl is 25°. The distance between the two end stars in the bowl is 5°. The sides of the Great Square of Pegasus average 15° in length. The distance from one end of the **W** of Cassiopeia to the other is 13°. The distance from Betelgeuse to Rigel in Orion is 19°, and the length of Orion's belt is just under 3°.

Your hand is a portable angle measurer. The width of your clenched fist held at arm's length is about 10° (people with shorter arms generally have smaller fists and the general rule holds). The width of a finger at arm's length is about 2°. The angular diameter of the Moon is only 1/2°, although most people would guess it is much larger. Try it yourself. Block out the full Moon with your little finger held at arm's length.

The *field of view* is an astronomical term that often confuses newcomers to astronomy. It is the angular width of the patch of sky you can see through your optical instrument, whether that instrument is the naked eye, a pair of binoculars or a telescope. Field of view is usually expressed in degrees for binoculars and *arcminutes* for telescopes. An arcminute is 1/60 of a degree. An even smaller unit, an *arcsecond*, is 1/60 of an arcminute.

The entire sky from horizon to horizon encompasses 180°, but we see only part of it at once, and with binoculars and telescopes we see very little at a time. With the naked eye, we have a field of view of about 100°, which is the field of view you see when you open Starry Night. Binoculars typically give fields of view of 5° to 7°, which is a large enough portion of the sky to see the bowl of the Big Dipper or the belt of Orion, but not an entire constellation. A telescope with a low-power eyepiece typically gives a field of view of 30 arcminutes (expressed as 30'), which is the equivalent of 1/2°. A high-power eyepiece on the same telescope may give a field of view of 10 arcminutes (or 1/6°). It is difficult to find objects with a high magnification telescope because you can see only a very small part of the sky at one time. Starry Night displays your field of view and can also display circular outlines which represent the fields of view of different optical instruments.

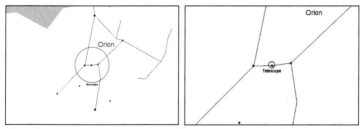

The relative views of Orion's belt as seen through 7 X 50 binoculars (7° field of view) and a low-power telescope eyepiece (30' field of view).

## Stars

Gaze upwards on a clear night and you see – stars! A lot of stars, if you are lucky enough to be in a dark place. How many is a lot? From a dark location a person with good eyesight can see about 2,000 stars at any moment, or about 6,000 if he could see the entire sky.

It is useful to have a system for classifying the brightness of stars, and the one we use has been around for thousands of years. When compiling his star catalog in about 150 BC, the astronomer Hipparchus devised the scheme still used today. He divided the naked-eye stars into six *magnitude* groups, with lower numbers designating brighter stars. First magnitude stars are the brightest and sixth the dimmest that most people can see without a telescope. It is counter-intuitive for the **brighter** stars to have a **smaller** magnitude number, but the scheme has been used for so long that astronomers have gotten used to it.

Modern astronomers have recalibrated Hipparchus' scheme to put it on a sound mathematical footing, and magnitudes are now expressed as decimals. The brightest star, Sirius, is now assigned a magnitude of -1.4. The planets can be even brighter, Jupiter reaching -2.9 and Venus -4.5. The full Moon is magnitude -12.6 and the Sun is -26.7. The magnitude of the dimmest object you can see is called the *limiting magnitude*. With the naked eye on a clear night, the limiting magnitude is about 6th magnitude, but with light pollution it can be much higher. Binoculars and small telescopes will see down to 9th magnitude, a large amateur telescope to 14th magnitude (the magnitude of Pluto), and the Hubble Space Telescope reaches to 30th magnitude. Starry Night can show you the magnitude of any object.

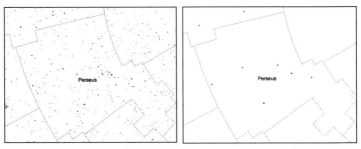

The stars of Perseus visible with a limiting magnitude of 6, and a limiting magnitude of 3.

## Astronomical Distances

Most people know that distances in outer space are quite large – "astronomical," so to speak. Our normal measurements of length become meaningless in these situations, and astronomers have invented two new units of length that serve different purposes in astronomy.

The *astronomical unit* (AU) is the average distance from Earth to the Sun. This equals 150 million kilometers (93 million miles). It is used primarily to describe distances to objects within the solar system. For example, Jupiter is about 5 AU from Earth on average, while Neptune is about 30 AU away.

When talking about the distances to the stars, even the astronomical unit is not large enough. The nearest stars are about 250 000 AU away, which is a remarkable 37,500,000,000,000 km ( 24,000,000,000,000 miles)! For distances to stars and other deep space objects, astronomers use the *light year*. A beam of light (or radio waves or any other sort of electromagnetic radiation) travels at "the speed of light," which is 299,792 kilometers (186,282 miles) per second. That is fast enough to travel seven times around Earth in one second or to the Moon and back in three seconds. At this incredible rate, a beam of light travels 9,460,540,000,000 kilometers (5,878,507,000,000 miles) in one year – and that distance is one light year. A light year is a unit of **distance**, not **time**, and it is approximately 9-1/2 trillion kilometers (6 trillion miles).

Each magnitude step represents a difference in brightness of 2-1/2 times; for example, a 3rd magnitude star is 2-1/2 times as bright as a 4th magnitude star. (The actual ratio is 2.512, which is the 5th root of 100; the difference between a 1st and 6th magnitude star is 100 times.) The Hubble Space Telescope can photograph stars 24 magnitudes or 4 billion times fainter than the dimmest stars your eye can see.

The magnitude of a star depends on two things: the amount of light it gives off and its distance from Earth. The stars that can be seen from Earth range in distance from about 4 light years to several thousand light years, making it impossible to determine how intrinsically bright a star is by its magnitude alone.

# Learning to See

Although sixth magnitude is the theoretical naked-eye limit for most people, you may have trouble seeing objects this faint at first. It takes practice to learn to sing a song, throw a baseball, or look at a small, faint object through a telescope – or at least it takes practice to do it well. You must learn how to see. Slow down, and let your eye absorb the image. Astronomical objects are small and contrast is low, so details do not spring out. The most important thing to do is to relax, linger at the eyepiece, and let an image slowly accumulate on your eye.

The greatest amateur astronomer of all and the discoverer of the planet Uranus, William Herschel, said, "You must not expect to see 'at sight.' Seeing is in some respects an art which must be learned. Many a night have I been practicing to see, and it would be strange if one did not acquire a certain dexterity by such constant practice." Go slow, and take the time to see.

# The Night Sky
# Starry Night Exercise

This exercise is designed to complement the concepts discussed in this chapter. You should open Starry Night on your computer before beginning the exercise.

As mentioned earlier, the sky is divided into different constellations. Starry Night lets you display the constellations in several different ways.

1) Open a new window by selecting "File I New."

2) Choose "Go I Viewing Location" to bring up the Viewing Location window.

3) Click the "Lookup" button and select Chicago from the list of cities. Click the "OK" button. The coordinates for Chicago should now appear in the Viewing Location window. Make sure that the box marked "DST" is not checked. Click the "Set Location" button to change your viewing location to Chicago.

4) Change the date to Dec.1, 2000 and change the time to 10:00 PM. Press the "Stop" button in the time controls to keep time fixed at 10:00 PM.

5) Select "Constellations I Boundaries" to draw the constellation boundaries on the screen and then select "Constellations I Labels" to label them. Select "Constellations I Constellation Settings" and press the "Labels" button in the Constellation Settings window. In the Label Settings window which comes up, make sure the option "Both" is selected (under the heading "Constellation Label Options"). This displays the common and astronomical names of the constellations. Press the "OK" button to close these two windows. Scroll around the sky using the mouse and notice the different sizes of the constellations and their irregular shapes.

6) Select "Constellations I Astronomical" to turn on the common stick figures. You can see that all the stick figures stay within their constellation boundaries. Again, scroll around the sky and note that few figures bear any resemblance to their names, although Orion (in the southeast) does resemble a hunter with bow, and the Great Bear (in the northeast) is shaped like an animal.

7) As an example of how you can use the constellations to look for an object, suppose you want to observe the star cluster M52. Open the Find dialog box by choosing "Edit I Find" and type in "M52" in the field marked "Name Contains." Make sure that the checkbox marked "Magnify for Best Viewing of Found Object" is not checked. Click the "Find" button to have Starry Night center and mark the object. You can see that M52 is in Cassiopeia. Select "Constellations I Labels" to turn off the labels, and notice the distinctive "W" figure which Cassiopeia makes in the sky. This "W" is one of the easiest star patterns to find in the night sky. You can see that the rightmost line of the "W" (the line that connects the stars Schedar and Caph) points to the cluster. This will help you point your binoculars on M52 when you go observing (note that M52 is too faint to see with the naked eye).

8) Now turn off all the constellation indicators. Select "Constellations I Boundaries" to turn off the boundaries, "Constellations I Labels" to turn off the labels, and "Constellations I Astronomical" to turn off the stick figures.

You learned in this chapter that another way to find an object is to know its coordinates in the Horizon system: its altitude and azimuth.

1) Starry Night shows you the altitude and azimuth of the **center** of the screen when you are scrolling around the sky. This is called your gaze information. The gaze information appears along the bottom center of the screen. To pull up the gaze information at any time, just hold down the left mouse button.

2) Remember that altitude measures how far above the horizon an object is located. Scroll up in the sky until the zenith (marked in red with two curved arrows when you are scrolling or holding down the left mouse button) is near the center of the screen. By looking at the gaze information, you will see that the altitude is near 90 degrees. Similarly, if you scroll down in the sky so that the horizon is near the center of the screen, you will notice that the altitude is near 0 degrees.

3) Azimuth is measured in degrees from North. If you scroll around so that the compass marker for North is near the center of the screen and then hold down the left mouse button, you will see that the azimuth is close to 0 degrees (or just less than 360 degrees – because the sky is divided into 360 degrees, the counter "resets" at 360). Likewise, if you move around on screen so that the East compass point is near the center of the screen, the azimuth will be near 90 degrees.

4) Now, we will use an object's Alt/Az coordinates to find it in the sky. You are told that Sirius has an altitude of about 10 degrees, and an azimuth of about 124 degrees (remember that Alt/Az coordinates are only valid for a specific time and will change as the star moves in the sky – these coordinates are valid for 10 PM on Dec.1, 2000 from Chicago). Use your understanding of how these coordinates relate to the horizon and the direction markers to locate Sirius, the brightest star in the sky. Remember that at any time, you can hold down the left mouse key to find the Alt/Az coordinates of the center of the screen.

5) Starry Night can also tell you the Alt/Az coordinates of an object at any time. Open the Find dialog box and find "Polaris," also known as the North Star. Starry Night will then center on Polaris and identify it. Double-click on Polaris to pull up its information window (if no window opens, you didn't have the cursor positioned directly over the star. You should see the cursor icon change from a hand to an arrow before you double-click on the star). Make sure the dropbox in this window reads "Local (Alt/Az)." Polaris's altitude and azimuth are shown just beneath the dropbox. Note that the altitude is approximately equal to your latitude (the latitude of Chicago is about 41.5 degrees N) and the azimuth is very close to 0 or 360 degrees. This is always the case for the North Star – it is the one star in the sky which doesn't appear to move over time, so its Alt/Az coordinates remain constant. Close the Info Window.

6) Select "Guides | Celestial Poles." This will turn on markers for the North and South Celestial Poles. The North Celestial Pole has an altitude exactly equal to Chicago's latitude, and an azimuth of 0. Note how close it is to Polaris. You may want to choose "Edit | Select None" to get rid of the label for Polaris, because it will overlap the label for the North Celestial Pole.

You will now use Starry Night to get a better understanding of the different angular sizes of objects in the sky. Starry Night shows you a field of view approximately equal to what you can see with your eyes, about 100 degrees. Your current field of view is always displayed near the two zoom buttons.

1) Open the Find dialog box and type in "Andromeda Galaxy." This will center on the Andromeda Galaxy. It is almost invisible in your current field of view. (Make sure the "Magnify for best viewing" checkbox is still turned off or the galaxy will fill the screen).

2) Select "Guides | Field of View Indicators | 7 X 50 Binoculars." This draws a circle over the center of the screen indicating the patch of sky you would see if you were looking through binoculars of this common type.

3) Use the zoom button to zoom in on the Andromeda Galaxy. Zoom in until the circular outline covers almost the entire screen. Notice how the field of view changes as you zoom in. Now you can see how the size of the galaxy compares to the field you see through your binoculars. The binoculars have a field of view of about 7 degrees across (the number shown near the zoom buttons will be a bit larger at about 10 degrees, because it is the angular width from the left side of the screen to the right side). The angular width of the Andromeda Galaxy is just under 2 degrees. You can see that the galaxy still appears quite small when seen through binoculars of this type.

4) Press the square button to the right of the zoom buttons. This restores your field of view to 100 degrees.

5) Select "Guides | Field of View Indicators | 7 X 50 Binoculars" to remove the circular outline from the screen.

To conclude this exercise, you will learn a bit more about magnitudes. When you are looking at a regular 100 degree field of view, Starry Night shows all stars brighter than magnitude 5.7, which matches approximately what you could see in a dark sky. As you zoom in, the limiting magnitude automatically increases to show you dimmer objects. If you hold down the left mouse button, the limiting magnitude is shown along the top of the screen.

1) Open the Find dialog box and type in "Sirius."

2) Bring up the Info Window for Sirius. The star's magnitude is shown in the window that appears. For Sirius, the magnitude is -1.47.

3) By bringing up Info Windows for other nearby stars, you can see how magnitude relates to brightness. For example, the three stars in Orion's belt (above Sirius, slightly to the right) all have magnitudes between 1.6 and 2.3, about 3 to 4 magnitudes dimmer than Sirius, while Betelgeuse (the bright star at the top of Orion) has a magnitude of 0.4, about 2 magnitudes dimmer than Sirius.

4)  Changing the limiting magnitude affects the number of stars you see. Select "Sky | Small City Light Pollution." The limiting magnitude is now 4.5. Note how many fewer stars are visible.

5)  Select "Sky | Large City Light Pollution." The limiting magnitude is now 3.0 and only the brightest stars are visible.

6)  Which of the three views do you think best matches the night sky you normally see? Select "Sky | No Light Pollution," then turn on the constellation names and stick figures. Scroll around the sky and find Ursa Minor, also known as the Little Dipper. The seven majors stars that make up the Little Dipper have a wide range of magnitudes. Polaris and Kochab are fairly bright with magnitudes around 2, while at the other extreme, Eta Ursae Minoris has a magnitude of about 5. The other 3 stars have magnitudes in between these extremes. Find the magnitudes of all six stars (by bringing up their Info Windows) and write them down. The next time you are observing, determine which ones you can see with the naked eye. Your limiting magnitude is somewhere between the magnitude of the dimmest star in the Little Dipper that you can see, and the magnitude of the brightest star in the Dipper that you cannot see. If you can see all the stars in the Little Dipper, count yourself lucky!

# MOTIONS OF THE EARTH

There is movement in the sky. Not only do the Sun, Moon, planets, comets, and asteroids move against the background of stars, but the sky itself moves. These changes happen because of the motions of the Earth as it spins on its axis and orbits the Sun. As the night passes and as the seasons change, we face different parts of the universe and see different stars and constellations.

## Rotation of the Sky

Many celestial motions are too slow to be noticed over so short a period as a night or even a month, but the nightly rotation of the sky happens on a scale that, with a little patience, we can experience while we gaze upward. The sky's rotation is shown dramatically in long time-exposure photographs centered on the North Star, which show the motion of stars as circular trails of different sizes centered on the sky's North Pole. We speak of the sky rotating overhead, although we know that it is the Earth that is turning. The Earth makes one rotation a day, spinning from west to east, which causes the sky to turn from east to west. We speak of the Sun rising in the morning, although we know that it is the Earth that turns towards it, making the Sun appear to rise above the horizon. The illusion is so convincing that it wasn't until the time of Copernicus in the 16th century that people accepted that the Earth does indeed turn on its axis.

Long time-exposure photo showing star trails that arc as the Earth spins.

## Proving that the Earth Spins

How would you demonstrate that the **Earth** spins rather than the **sky**? No simple visual demonstration existed until the Frenchman Jean Foucault hung a massive iron ball from the high dome of the Pantheon in Paris in 1851 and set it swinging. Foucault demonstrated that this pendulum appears to slowly change the direction of its swing relative to the ground beneath. Since the pendulum does not feel the orientation of the building it is attached to, the Earth is free to rotate under it without affecting the direction of its swing. The pendulum feels the sum of the gravitational pull of the rest of the universe and maintains a constant orientation relative to the distant stars. Foucault Pendulums are found today in planetariums and science museums.

# Annual Motion of the Sun

The daily *rotation* of the Earth on its axis is one fundamental motion of the Earth (and of the sky). The second is the annual *revolution* of the Earth around the Sun.

Until the 16th century, it was taken as a matter of faith that the Earth does not move and is the center of all creation. Ancient Greek musings contrary to this view were taken as mere philosophical speculations. In 1543 the Polish astronomer Copernicus proposed that the Earth orbits the Sun, rather than the other way around, but he had no proof of what was to him a mathematical issue. Two generations later the great Italian astronomer Galileo Galilei supplied this proof in the form of telescopic observations of the phases of Venus and the moons of Jupiter. He took up the "heliocentric" (sun-centered) cause, but ran afoul of the authorities for his methods. The truth was out, however, and by the mid-1600s it was universally accepted (in Europe, at least) that the Earth orbits the Sun.

Long before anyone knew whether it was the Sun or the Earth that moved, astronomers plotted the apparent path of the Sun in the sky, relative to the background stars. This path is known as the *ecliptic*, and it can be displayed in Starry Night. Astronomers also noticed how the rise and set points of the Sun on the horizon and its noon-time elevation varied with the changing seasons. They even determined the length of the year – sometimes with surprising accuracy.

# The Zodiac

The Sun's path among the stars has been considered special since it was first identified. The Sun moves through certain constellations, and even in the earliest times these constellations were accorded extra importance. The Moon and planets pass through the

The Moon and planets all stay close to the ecliptic line.

same constellations (plus several others – see "The Extended Constellations of the Zodiac" in Chapter 3.3), and this also contributed to their mystique. Although you cannot **see** the Sun's path through the stars when you stand outside, Starry Night can show it to you.

The Moon stays close to the Sun's path and provides a simple way to divide it into segments. The Moon travels around the sky a little over 12 times while the Sun travels around it once (this is the long way of saying there are 12 months in one year; see Chapter 4.1). Rounding this to the convenient whole number 12 suggests that the Sun's (and moon's) path be divided into 12 segments, each of an equal length (30°). Doing this links the motions of the Sun and Moon at least symbolically.

Along the Sun's path are prominent groups of stars, like Scorpius and Gemini, and areas devoid of bright stars, like Aquarius and Cancer. The Sun passes through 13 of the 88 constellations mentioned earlier in its yearly journey through the stars. 12 of these 13 constellations are the classical constellations of the *zodiac*. They were all named by 600 BC, but most are far older. Scorpius, for example, has been seen as a scorpion for at least 6,000 years, which is long before the concept of the zodiac as a complete circle was worked out. Most people associate the zodiac with astrology.

## The 13th Constellation of the Zodiac

The constellation boundaries were arbitrary until recently, and each astronomer (and astrologer) was free to place the boundary lines where he saw fit. This caused endless confusion until 1930, when the constellation boundaries were fixed – among astronomers at least – by international agreement. (See "The Official Cygnus" in Chapter 5.2) One effect of this tidying-up was to draw the huge and ancient constellation Ophiuchus so that it intersected the ecliptic. The Sun is actually within the boundaries of Ophiuchus from approximately November 30 to December 17 each year. In contrast, the Sun is within Scorpius for only one week.

# Annual Changes in the Stars

The Sun's apparent motion against the background of stars also causes seasonal changes to the constellations that we see at night. Each day the Sun is nearly 1° to the east, relative to the stars, of where it was the day before. If we think of the Sun as staying relatively still (and - after all - our timekeeping methods are based on the position of the Sun, rather than the stars), we can think of the stars as moving westward 1° per day relative to the Sun.

Stars rise four minutes earlier each day, or 1/2 hour earlier each week, or 2 hours earlier each month, or 24 hours earlier each year. This is another way of saying that the cycle has been completed and the stars rise at the same time again after one year has passed. If a star rises at 2 A.M. on one date, it will rise at midnight one month later, at 10 P.M. another month later, and at 8 P.M. yet another month later (some stars near the North Star are exceptions to this rule, for they are *circumpolar*, meaning that they do not rise and set, but remain above the horizon all day and night). This is a handy rule of thumb to remember when you are planning which stars and constellations to observe. If you have to stay up too late to see it now, wait a few months and it will be conveniently placed in the evening sky. This rule of thumb does not work for the Moon or for Mercury, Venus, and Mars, as they have their own motion against the stars. It is relatively accurate as a rule of thumb for Jupiter and Saturn (and the outermost telescopic planets) and for most asteroids, because they orbit the Sun very slowly and thus appear to share the same motion as the stars.

Of course, the stars also **set** four minutes **earlier** each day, 1/2 hour earlier each week, and 2 hours earlier each month. If Saturn or Orion sets at 8 P.M. this month, it will set at 6 P.M. next month – and you won't see it. "What the Sun giveth, the Sun taketh away."

## Motions of the Earth
## Starry Night Exercise

This exercise will help you understand how the rotation of the Earth and the revolution of the Earth around the Sun alter the appearance of the sky. You should open Starry Night on your computer before beginning.

The Earth's rotation on its axis is responsible for the rapid changes in the position of the Sun, planets and stars in the sky.

1) Open a new window by choosing "File I New."
2) Select "Sky I Daylight" to turn sunlight off.
3) Choose "Go I Viewing Location" to bring up the Viewing Location window. Click the "Lookup" button and select Toronto from the list of locations. Click "OK." Make sure that the box "DST" is not checked and then press the "Set Location" button.
4) Set the time to 10:00 PM and the date to October 6, 1999. Scroll around using the mouse so that you are looking north instead of south.
5) Open the "Find" dialog and type "Polaris" to center on the North Star.

The Big Dipper is to the bottom left of Polaris, with its bowl almost parallel with the horizon. Note how the two "pointer stars" on the end of the Big Dipper's bowl form a straight line that points to Polaris (make sure that your Starry Night window is full size so that you can see the pointer stars and Polaris onscreen at the same time).

6) Set the time step in the time controls to 3 minutes, if it is not already set to this value. Now press the Forward button to start time running. You can see that the stars appear to circle counter – clockwise around Polaris. Watch how the Big Dipper changes its orientation relative to the North Star.

7) Notice how the speed at which a star moves is determined by its angular separation from Polaris. The farther from Polaris a star is, the more rapidly it moves in the sky. Stop the time by pressing the Stop button in the time controls.

This example shows you how the position of stars changes with the seasons. We say that Orion is a "winter constellation" in the Northern Hemisphere. Why is that?

1) Set the date to Dec. 21, 1999, and the time to 8:00 PM.

2) Open the "Find" dialog and type "Betelgeuse" (watch the spelling!). Betelgeuse is the bright star in Orion's shoulder (note the 3 bright stars to its right that make up the belt). Betelgeuse has recently risen above the horizon.

3) Right-click (click and hold on a Macintosh) on Betelgeuse to bring up a contextual menu. Select "Center\Lock" from this menu

4) Now double-click on Betelegeuse to open its Info Window. Write down its rise and set times and then close this window.

5) Change the time step from 3 minutes to 1 day (choose the step marked "day," not the one marked "day (sidereal)." A sidereal day is a different unit of time which is explained in Chapter 3.1).

6) Use the Single Step Forward button to move time ahead by one day. Betelgeuse is now farther up and to the right relative to the horizon. Double-click on it again to find its new rise and set times. You should find that they are both about 4 minutes earlier than the previous times. As you learned earlier, this is due to the Earth's revolution about the Sun.

7) Click the Forward (or Play) button to watch Betelgeuse's location in the sky change continuously. Stop time just before it falls beneath the horizon. What is the date? You should find that it is near the beginning of June.

8)   Click on the small outline of the Sun shown beside the time in the Control Panel. This should color the Sun yellow, meaning that Daylight Savings Time has been turned on.

9)   Double-click on Betelgeuse to find its new rise and set times. You should notice that its rise and set times mean that it is now above the horizon only during the day, so it will be invisible to us (remember that we turned off daylight for this example). But if you compare the rise and set times with those you found for Dec.21, you will notice that the total time which the star is above the horizon is exactly the same: 13 hours and 2 minutes! So a star or constellation is classed as "winter" or "summer" not because it is above the horizon longer at those times, but because it is above the horizon at night, when we can see it.

10)  Now change the time to Dec. 21, 2000. Click on the small outline of the Sun to turn off Daylight Savings Time. Double-click on Betelgeuse to find its rise and set times. You should find that they are almost the same as a year ago, as the Earth has returned to its previous place in the solar system relative to the Sun.

To close this exercise, we will do something you can never do in real life: watch the Sun move through the constellations of the Zodiac. Remember that this is caused by the Earth's revolution around the Sun.

1)   Set the time to 12:00 PM. We want to follow the Sun over the course of a year, so we have to choose a time of day when it will always be above the horizon.

2)   Scroll around the sky to find the Sun, and then right-click (click and hold on the Mac) on it to bring up a contextual menu. Choose "Center/ Lock" from the menu.

3)   Turn on the Ecliptic line by choosing "Guides | The Ecliptic."

4)   Now turn on the Zodiac constellations by selecting "Constellations | Zodiac." Turn on the constellation labels and boundaries as well. The Sun should be in Sagittarius.

5)   Make sure that the time step is still set to 1 day. Start time running by pressing the Forward (or Play) button in the time controls. If you follow the motion for a year, you can see how the Sun moves through all the constellations of the Zodiac, as well as moving through a part of the sky not in the classical Zodiac.

6)   Stop time when the Sun is outside the boundaries of the Zodiac. Place the cursor over the Sun. The screen will display the name of the Sun and the constellation it is in, which you will see is Ophiuchus.

# THE SOLAR SYSTEM

## Motion of the Moon

Is the Moon out tonight? Very likely – it is half the nights. If not, it is probably out during the daytime. Many people are surprised to see the Moon during the day, but it is visible in the day for two weeks each month. Only for about three days each month when the Moon is new or nearly new can it not be seen at some time of the day or night.

The Moon orbits the Earth in counterclockwise direction as seen from above the Earth's north pole. We see it move night by night across the sky from west to east. Each evening it is about 1/30 of 360°, or 12°, east of where it was the night before. It takes about 30 days to move through the constellations of the zodiac, and it spends about 2-1/2 days in each, on the average, before moving on to the next.

In addition to moving around the sky, the Moon changes its shape. Planets move too, but the Moon and rare bright comets are the only naked-eye objects in the sky that change their shape day by day. In the case of the Moon, the change is quite predictable.

The chief principle is that the illumination that strikes the Moon comes only from the Sun, and the Sun lights up the side of the Moon that faces it. The Sun lights up exactly one half of the Moon at any moment – but that is true for any ball lit by the Sun. Hold a tennis ball in sunlight, and no matter how you turn or position it, the half facing the Sun is the half that is lit. As the Moon orbits the Earth, the side facing the Sun remains lit. When the Moon is *new*, the side facing the Sun faces away from the Earth, and we see the dark side of the Moon. This is **not** the same as the back side, which is the side facing away from the Earth; the Moon's back side and dark side coincide only at full Moon.

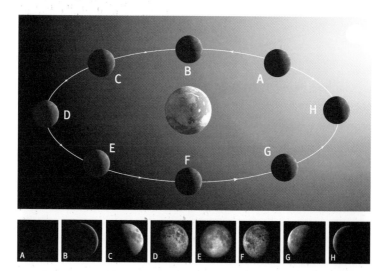

The diagram above shows the Moon at eight positions on its orbit, along with a picture of what the Moon looks like at each position as seen from Earth.

As the Moon moves out of alignment with the Sun, it appears to the east of the Sun in our sky, and it sets after the Sun. We now see most of the dark side, as before, but we begin to see a thin crescent of the lit side – we see a crescent Moon in the evening sky. One week after it is new, the Moon has traveled one-quarter of the way around the Earth and around the sky – and now it looks like a half-Moon. We see half of the lit side and half of the dark. After another week the Moon has moved another quarter of the way around the Earth and is now opposite the Sun, and we call it a *full Moon*. We are between the Moon and Sun and we see the Moon's lit side. Now – and only now – is the Moon's back side the same as its dark side. After another week, the Moon has moved three-quarters of the way around the Earth and is in a similar position to when it was at first-quarter, but now the **left** side of the Moon is illuminated, rather than the **right**. It looks like a half Moon. During the final week of the lunar month it is a crescent again that grows progressively slimmer.

During the period between first and last quarter, the Moon's phase is called gibbous. It is a *waxing gibbous* between first-quarter and full and a *waning gibbous* between full and third quarter (waxing and waning are from Old English and mean "increasing" and "decreasing" respectively). From new to first quarter it is a *waxing crescent*, and between third quarter and new a *waning crescent*.

# Eclipses

A *solar eclipse* happens when the Moon moves in front of the Sun and blocks the Sun's light. The eclipse can be either partial (the Moon does not completely cover the Sun and blocks only part of it) or total (the Moon completely covers the Sun). A partial eclipse can be *annular* if the Moon moves directly in front of the Sun but is not large enough to completely cover it, leaving a ring of sunlight ("annular" comes from the Latin word for ring) shining around the edge of the Moon. Annular eclipses exist because the orbit of the Moon around the Earth is an ellipse, not a circle. This means that the Moon is not always the same distance from Earth. When it is farther from Earth than normal, it does not appear as large in the sky. Therefore, it cannot completely cover the Sun and we see an annular eclipse.

Only during a total eclipse does the sky grow dark and the Sun's dazzling corona burst forth in radiant glory. By an accident of the Earth's history, our Sun is about 400 times larger than our Moon, but also 400 times farther away. This means that both objects

## Observing Solar Eclipses

The Sun's surface is too bright to look at without risking serious eye damage, and this remains true even when most of the surface is covered by the Moon. Always use proper precautions when observing the Sun, both when watching an eclipse and when looking for sunspots. Use an approved dark or mirrored filter placed in front of your eye or in front of your telescope's main lens or mirror (do not use a filter that screws into your eyepiece) or project the Sun's image onto a white surface. Do not use home-made filters that may block the Sun's visible light but allow harmful amounts of infrared or ultraviolet light to pass through. You can purchase solar filters from a variety of sources.

Only during the brief totality phase of a total eclipse is it OK to look at the Sun without filters – and you should, to enjoy the spectacle!

appear to be almost exactly the same size. For this reason, total eclipses are very brief, because the Moon and Sun have to be lined up almost exactly, and the motion of the Moon prevents a total eclipse from lasting more than a few minutes.

During each eclipse, a similar sequence of events happens. Open the file "SunEclipse1" to follow the course of a partial eclipse as seen from Cape Town, South Africa, in April 2004. The Moon appears as a pale disk near the Sun. The Sun will appear to stand still while the Moon approaches, eclipses it, and moves on. First contact is the moment when the Moon's edge first "touches" the Sun and the eclipse begins – in this eclipse it is at 2:54 P.M. Second contact is when the Moon leaves the Sun and the eclipse ends; in this case it is at 5:23 P.M. As seen from Cape Town, the Moon never covers more than half of the Sun's face.

For total and annular eclipses there are four stages. Open the file "SunEclipse2" (an eclipse that will be widely visible in the far east in 2009). First contact is at 8:24 A.M. Second contact now becomes the moment when the Moon first completely covers the Sun; it is at 9:37 A.M. The eclipse is total between second and third contacts, and totality lasts only a few minutes (about 5 minutes for this eclipse). The eclipse ends at fourth contact (11:01 A.M.). You may wish to view the moment of totality again, in slower motion. Stop time and change your time step to 1 second. Then change the time to about 9:35 A.M. before starting time running forward again.

Annular eclipses follow a similar sequence except that between second and third contacts the Moon is silhouetted against the Sun, which appears as an unbroken ring for a few minutes. Open the file "SunEclipse3" to view an annular eclipse in the spring of 2005. Notice how the sky does not darken completely, even though the

Sun is almost completely covered by the Moon at the eclipse's peak.

The precise times and circumstances of a solar eclipse depend on your observing location. You stand in the shadow of the Moon during a solar eclipse, and the central part of the Moon's shadow (the *umbra*) is no more than 269 km (167 miles) wide when it reaches the Earth. If you are within it, the eclipse is total, but if you are not, it is partial or you miss it entirely.

The Moon is moving and so is its shadow. The Moon's shadow sweeps over the Earth at about 1,600 km (1,000 miles) per hour, racing from west to east. Where the shadow first strikes the edge of the Earth, observers see the Sun (which is directly behind the Moon) low in the east; it is a sunrise eclipse. People at the center of the shadow experience the eclipse at noon, and people at the eastern edge of the Earth see it at sunset. Although the path of totality (or

## A Solar Eclipse From the Moon

People on the Moon during a total solar eclipse would see nothing unusual happen to the Sun, but they would see the shadow of the Moon race across the Earth. It would resemble a lunar eclipse as seen from Earth, but the Moon's shadow is far smaller than the Earth's shadow and it appears as a dark splotch with a black center that moves across the surface of the Earth.

Open the file "SunEclipse4" to see the July 2009 eclipse from the Moon. Everywhere within the area of the large shadow experiences a partial eclipse while the tiny black dot in the middle of the shadow marks the path of totality. Zoom in if you want a closer look at the area of totality.

annularity) is narrow and comparatively few people see a total or annular eclipse, the outer part of the Moon's shadow is wide and much of the Earth's population sees a partial eclipse at the same time.

The solar eclipse has a less spectacular counterpart in the lunar eclipse, caused by the Earth passing between the Sun and the Moon. Lunar eclipses are described in more detail in Section 4.1.

# Planets

While the stars appear in the same configuration relative to each other every night, the planets do not. Much like the Sun, they move slowly through the sky from constellation to constellation. Some planets move much faster than others. All the planets move through the same constellations – more or less – as the Sun. This is because the planets all move around the Sun in nearly the same plane, orbiting above the Sun's equator. From Earth, we see this plane edge-on and it appears to us as a line called the ecliptic. The Sun always remains on this line and the planets never stray far from it.

Although the Sun, Moon, and planets move across the sky and it appears that we are at the center of all we see, we know now that this is not the case. The Sun is at the center of our solar system, and the planets (including the Earth) circle around it counterclockwise as seen from above the Sun's north pole. The planets all move in the same direction, but not at the same speeds. The closer a planet is to the Sun, the faster it moves because it feels the pull of the Sun's gravity more strongly. Planets near the Sun also have shorter paths and complete their years more quickly; Mercury orbits the Sun in only 88 Earth days but Pluto takes 248 Earth years. Each planet travels at a nearly uniform speed on a nearly circular orbit. (Mercury and Pluto are the only noticeable exceptions to these last two, but they follow the general rule of distance by moving fastest when closest to the Sun.)

Perhaps the most profound observation – and one that would have astounded all ancient people – is that **the Earth is one of the planets and moves with and among the others**. We view the sky from a platform that is itself in motion.

# The Solar System
# Starry Night Exercise

After working through this exercise, you will know how to use Starry Night to learn more about the motions of the Moon and planets.

We begin by watching the motion of the Moon.
1) Open a new window by choosing "File | New."
2) Turn off daylight by selecting "Sky | Daylight."
3) If it is not already open, open the Planet List by selecting "Window | Planet List."
4) Click on the name of the Moon in the Planet List, and then click the Lock icon. This should center the Moon. If a window comes up saying the Moon is beneath the horizon, click the "Reset Time" button in this window.
5) Two fields in the Info Window can tell you about the Moon's age. Double-click on the Moon to open the Info Window. "Disc Illumination" tells you what percentage of the Moon's face is illuminated, ranging from 0% at new Moon to 100% at full Moon. "Age" tells you how long it has been since the Moon was last new, and also gives the phase name. The Moon rise and set times are also shown. Close the Info Window.
6) Turn on the Zodiac constellations and the Ecliptic line.

7) Change the timestep to 1 day.

8) Press the Single Step Forward button to watch how much the Moon moves in a day. Keep pressing the Single Step Forward button to watch the phases change (if the Moon falls beneath the horizon, continue clicking the Single Step Forward button until it reappears). Double-click on the Moon at any time to find out its age and phase. Note how the Moon stays close to the ecliptic, always remaining within about 5 degrees.

We will move on to study the motion of the planets.

1) Select "Labels | Planets/Sun" to turn on labels for all the planets. Scroll around the sky and notice how all planets above the horizon are near the line of the Ecliptic.

2) In the Planet List, click on Jupiter's name and then click the Lock icon. If Jupiter is beneath the horizon, click the "Reset Time" button in the window which comes up.

3) Start time flowing by pressing the Forward (or Play) button (if Jupiter falls beneath the horizon, just continue to let time flow until Jupiter comes back above the horizon). You will see that Jupiter also moves through the Zodiac constellations, but much slower than the Sun did. To see why this is so, we need to change our viewpoint to see the motion of the planets from a new perspective.

4) Stop the time.

5) Choose "Go | Outer Solar System" from the menu. This places you high above the Sun, looking back at the solar system. The orbits of all the planets are shown. The orbit of Jupiter is in red.

6) Use the zoom buttons to zoom in until the orbit of Jupiter almost fills the screen. You will now be able to see the orbits of the inner planets, including Earth.

7) Change the time step from 20 days to 10 days. Press the Forward button to start time running. You can see how quickly the inner planets move compared with Jupiter. The outer planets of Saturn, Uranus, Neptune and Pluto move even slower (which you can see for yourself by zooming a little farther out and using the zoom button. This is why Jupiter moves through the Zodiac so slowly.

This concludes the exercise for this chapter.

# Section 2
# Observational Advice

S TARRY NIGHT is a great tool for learning more about astronomy, but it will never replace stepping outside and observing the wonders of nature with your own eyes. This section gives you some help on getting more out of your observations. Chapter 2.1 gives general hints & tips, as well as brief advice on buying binoculars or a telescope. Chapter 2.2 explains the different astronomical coordinate systems, which you will use to help you locate objects in the sky.

# SKYWATCHING

## Hints & Tips

Although casual skywatching takes no special training or skills, there are some things to keep in mind that will improve the experience.

First, as often as possible, observe away from city lights. Through all of history – until this century – the sky was dark at night even in the world's major cities, and (ignoring chimney smoke) the stars shone as brightly over New York City as over a rural Kansas town. Our modern cities are bathed in a perpetual twilight from the millions of lights that illuminate the air itself, not to mention any dust, smoke, and smog in the air, and this lets us see only the brighter stars. Astronomers refer to this unwanted stray light as *light pollution* and some campaign against inefficient lighting. Children who grow up in the city do not know what the stars look like and have no appreciation for the beauty of the sky, and even adults often forget. It is common to hear an audience gasp and even applaud when the stars come out in an urban planetarium show.

People are most aware of the poor quality of an urban sky when a bright comet appears and they are told by the news media to "go to a dark site" to see it properly. Even if a comet can be seen from within a city, it is a pale shadow of what people see from a dark location.

Many people conduct their most rewarding stargazing while on vacation. The national parks and national forests are generally exceptionally dark at night, and the sky becomes an attraction once darkness has settled over the trees, meadows, and waterfalls. Take binoculars or a spotting telescope on your next vacation and continue to sightsee after sunset.

A different type of light pollution is caused by our Moon. Even the darkest locations suffer when the Moon is bright. For two weeks each month centered on the date of the full Moon, moonlight bathes the sky in its own whitish twilight, reduces contrast, and obscures faint objects. There is nothing one can do about moonlight other than to try to work around it. During the week before full Moon, the Moon sets before sunrise and the sky is fully dark only before morning twilight begins; during the week after full Moon, the sky is dark briefly between the end of evening twilight and moonrise.

Another observing consideration is comfort. In most places it is cold in winter, especially at night, and winter stargazing means keeping warm. It is one thing to shovel snow or walk briskly to the store on a cold day; it is quite another to stand motionless and watch Orion in the dark. Your body generates little heat while watching the stars, and most people must dress warmer than anticipated. This is often true even in the summer, when it cools off at night. Padded footwear is important in the winter if you will be standing on cold ground. Many people watch meteor showers from the comfort of a sleeping bag on a lawn chair or ground pad. Bugs can be a problem too. Bug spray is an essential accessory for many summer stargazers (but be sure to not get spray on your optics!).

Another useful skywatching accessory is a red flashlight. Astronomy specialty shops sell observers' flashlights that shine with a very pure red light, but you can adapt a conventional flashlight by fitting a few pieces of red gel purchased at an art supply store over the lens. If you want to run Starry Night outdoors on a laptop during your observing sessions, you may also wish to make a red gel cover to fit over your computer screen.

Using a red flashlight preserves your ***night vision***. Step outdoors from a brightly lit room and notice how long it takes your eyes to adapt to the dark. After a few minutes your pupils will adjust to the lower light levels, and you will be able to see more stars. Turn on a white flashlight and look at a printed star chart. When you look up at the sky again, you will notice that your enhanced night vision is gone. Once you have regained it, look at the same printed chart with a red flashlight and then look back at the sky to appreciate the improvement. Using red lights to preserve night vision is a familiar trick to submariners and others who may have to suddenly leave a lit room and step outdoors ready to see in the dark.

## Binoculars and Telescopes

Your eyes are a perfectly fine tool for stargazing, and there is much to see with them alone on a dark night. But all sky watchers yearn to see more and fainter objects and to see them better, and eventually they want to buy a telescope or pair of binoculars. At the Griffith Observatory in Los Angeles, we recommend that people on a budget (or those who are very concerned with portability) start with a pair of binoculars. Binoculars are under-appreciated and are wonderful for many kinds of casual stargazing, and even experienced observers enjoy having a pair of good binoculars for wide-field views of the Milky Way (and for their portability!). You shouldn't leave home without a pair.

Binoculars have the advantages of being relatively inexpensive, highly portable, easy to use, and they give the best views of the Milky Way (and often of bright comets). Plus, you can use them for non-astronomical sightseeing, especially while on vacation.

Expense is relative, and in general you get what you pay for. Do not be tempted by $50 binoculars at discount camera stores and drug stores. They use plastic parts (including lenses!), are not well assembled, and should only be reserved for situations where they are at risk (such as boating, or perhaps for use by an accident-prone child). Anticipate spending $150 or more for a decent pair that will give good color-corrected images and that will provide a lifetime of use and enjoyment. A good pair of binoculars can be enjoyed for decades and handed down to the next generation.

Astronomers have different requirements than sportsmen, and their choice of binoculars is somewhat different. Adequate light is seldom a problem at the horse races, but it certainly is when peering at the Milky Way, so astronomers want binoculars with large lenses. The lens size is expressed as its diameter in millimeters, and it is the second number that characterizes a pair of binoculars. Astronomers want 50 mm lenses (or larger), and they use smaller binoculars only when size and weight is critical, as when backpacking. Larger binoculars with 70 or 80 mm lenses are great for astronomy, but they are expensive and they must be used on a tripod. The other characteristic of binoculars is the magnification, which appears as the first number. Magnification is not so important, and in any case comes within the range of 7 to 12. 10-power is a popular compromise, although it is hard to hold them steady without a tripod.

## Light Collecting Area

A 50mm binocular lens does not sound much larger than a 35mm lens, and a 20cm mirror does not sound much larger than a 15cm mirror, but in both cases the larger is **twice** the size of the smaller — if you are comparing the light-collecting area. A large lens or mirror collects more light than a small one and reveals fainter objects, and this is vitally important in astronomy. Compare the surface area, not the diameter. The area of a circle is proportional to the length of the diameter squared (the diameter times itself), so a small increase in diameter makes a big difference in area. The ratio of the lens area of 50mm and 35mm binoculars is $50^2/35^2$ = 2500/1225, or two to one; similarly, the ratio of a 20cm to a 15cm telescope mirror is 400/225 or almost two to one.

The best all-purpose binoculars for stargazers, then, are 7X50 to 10X50; seven to ten power with a lens diameter of 50 mm.

Test before buying. If you wear eyeglasses to correct for distance but have no significant astigmatism, use binoculars without glasses. Check before buying that they will focus at infinity to your sight with your glasses off. You probably won't be able to focus on an object which is far enough away while you are inside a store. Ask to test them outside and focus on a distant tree or building. Some people who normally wear glasses use contact lenses when observing, especially to avoid problems when switching back and forth between the sky, a finder scope, binoculars, the eyepiece, and star charts.

One step up from binoculars – and a good starter telescope – is a "spotting telescope." These telescopes are easy to use and feature low magnification and a wide field of view. While standard binoculars have a magnification of 12X or less, a spotting telescope will reach 40 power or more, giving great views of the Moon and bringing in the moons of Jupiter, the rings of Saturn, and many double stars. They're great for watching wildlife as well. Their name comes from their use of "spotting" the hits in target practice, but many amateur astronomers have a spotting telescope for vacations and travel. Prices range from $100 to well over a $1000 for a 70 – 100 mm spotting telescope with eyepieces, tripod, and carrying case or bag.

Recommendations for purchasing an astronomical telescope are beyond the scope of this book, but here are a few hints.

Amateur telescopes come in three flavors: refractors, which use a lens; reflectors, which use a mirror; and hybrids called Schmidt-Cassegrains. All telescopes also use eyepieces to magnify images.

Refractors are easier to make in small sizes and they dominate the market in the lowest price range; they are what you see in toy stores. Reflectors are more complex and are seldom sold for less than $200. Larger refractors with a lens 8 cm (3 inches) in diameter or larger can be superb instruments, but they are more expensive than a reflector of the same size and are starter telescopes only for the rich and famous. Reflecting telescopes with mirrors in the 15 to 20 cm (6- to 8-inch) range are the workhorses of amateur astronomers.

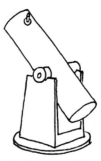

Reflecting telescopes with the Dobsonian-style mounting have become popular recently and are a good buy with prices starting at about $500; they are also relatively easy for hobbyists to build using common workshop tools and following published plans. Later you can work up to (or build) telescopes with apertures of 14

**Dobsonian-style mounting**

to 20 inches or more! Schmidt-Cassegrain telescopes are compact and portable but more expensive for a given size than simple reflectors.

Know where to shop – and where not to shop. Avoid discount camera stores (the kind that are always going out of business), department stores, and toy stores. Stay away from telescopes promoted on the basis of their magnification. A "464-power!" wonder is certainly a case of false advertising at best. To avoid disappointment, patronize telescope stores (which unfortunately are few and far between), the better camera stores, and catalogs.

Do your homework before you put your money down. Read the magazines *Sky & Telescope* and *Astronomy* (sold at major newsstands). They regularly print review articles on telescopes and other equipment for amateur astronomers, and their advertisers are a great source of free information. Request, read, and compare catalogs. Mail-order telescope sales are big business and make sense for people who live far from cities with proper telescope stores. Find out if there is a local astronomy club, and if so, attend a meeting. The club's members will be a wonderful source of practical information and recommendations. Clubs are also a source of good used equipment. A local planetarium or science museum can also offer advice. Some have large gift shops and sell telescopes and accessories. Last but not least, there are many excellent Internet resources to help you with your search. The major telescope manufacturers maintain sites with extensive information on their products, and there are also many independent sites which compare the merits of different telescopes.

No matter what type of telescope you choose, it is good to have realistic expectations of what you will see through your new instrument. It won't perform like the Hubble Space Telescope! You won't see the colorful images found in astronomy books and magazines; such photographs are long time-exposures taken with professional telescopes. But you will see the real, genuine thing – and that personal experience is worth a lot.

# COORDINATE SYSTEMS

Whether you are looking through your new telescope or just gazing up with the naked eye, you'll need to have a good understanding of the different astronomical coordinate systems to find many of the more interesting jewels in the night sky.

What is north? ... east? ... south? ... west? These directions are astronomical. The sky's rotation defines directions on the ground, positions on the Earth, and the main astronomical coordinate system used to specify positions in the sky. In Section 1, the Altitude-Azimuth coordinate system was described. We saw that any object in the sky could be uniquely identified by its angular height above (or below) the horizon and its compass direction. This is called a "Horizon" or "Local" coordinate system because an object's coordinates in this system are valid only for one location and only for one time. Astronomers use other more general coordinate systems to describe the positions of objects in the sky.

Of course, it is the Earth's rotation that determines directions and coordinates, but when the coordinate system was set up more than 2100 years ago it was almost universally "known" that the Earth is spherical but stationary while the sky turns overhead. The person who invented the system of using longitude and latitude to specify positions on the Earth was the Greek astronomer and mathematician Hipparchus. He rejected the notion that the Earth rotates on its axis, but his solution works just the same.

The Earth's rotation determines the position of the Earth's two poles and equator. The poles are where the Earth's rotation axis penetrates the Earth's surface as it continues from the Earth's center infinitely far into space, and the equator is the line midway between the poles. Lines of *longitude*, called *meridians*, run from pole to pole and divide the Earth into segments that specify a point's distance east of the zero segment, which last century was arbitrarily chosen to be the longitude of the Greenwich Observatory in England (the Prime Meridian). Longitude is measured either in degrees, in which case there are 360° around the Earth's circumference, or in hours, minutes, and seconds, which represents the actual rotation of the Earth. If you live 7 hours west of Greenwich, a star is overhead at your location 7 hours after it is overhead at Greenwich. There are 15° in one hour of longitude, which is the amount the Earth rotates in one hour of time. Hours are subdivided into minutes and seconds, and as you learned in Chapter 1.1, degrees are subdivided into arcminutes and arcseconds.

Longitudes in Starry Night are expressed in degrees east or

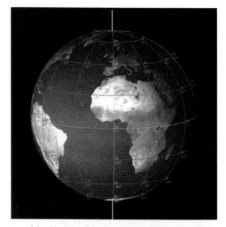

**Earth's lines of latitude and longitude.**

west of Greenwich, while *latitude* is the angular distance of a location north or south of the Earth's equator, expressed in degrees. The latitude of the equator is 0°, of the North Pole 90° N or +90°, and of the South Pole 90° S or -90°.

Positions on Earth are expressed as their longitude and latitude, in that order, which is the same as the angular distance west (or east) of the Prime Meridian passing through Greenwich and the angular distance north or south of the equator.

There is a corresponding system which astronomers universally use to specify positions on the *celestial sphere*. It is called the *equatorial coordinate system*. The celestial sphere is a concept, not a real object, but it is convenient to think of the globe of the sky as an invisible crystalline sphere that has the stars attached to it, with the Sun, Moon, and planets moving along its surface (it is an actual aluminum sphere in a planetarium theater). Just as positions on Earth are expressed by:

1) the time it takes the Earth to rotate from the zero meridian at Greenwich to the specified location,

2) the angular distance north or south of the Earth's equator, so celestial positions are expressed in terms of:

1) the time the sky takes to rotate from an arbitrary zero meridian to the specified point,

2) the angular distance north or south of the sky's celestial equator.

The two values are called *right ascension* and *declination* respectively, abbreviated RA and Dec. Values in right ascension are expressed in hours, minutes, and seconds; values in declination are expressed in degrees, arcminutes, and arcseconds. For example, the equatorial coordinates of Rigel would normally be expressed in this format: 5h 14m, -8° 12'.

To determine an object's right ascension, we first need to choose a line of 0h RA. Much like the line of 0° longitude on Earth is the

Prime Meridian, the line of 0h Right Ascension is known as the Celestial Meridian. This should not be confused with the local meridian, which is an imaginary line in the sky running from north to south through the zenith. Which stars the local meridian passes through depends on your location, whereas the celestial meridian always passes through the same stars and constellations.

Defining the celestial meridian is a two-step process: we first define the *vernal equinox* as one of the two points on the celestial sphere where the celestial equator meets the ecliptic (the vernal equinox is also the **moment** in time when the Sun crosses the celestial equator in March; it is the moment when spring begins in the Earth's northern hemisphere – see Chapter 3.2). The *celestial equator* is the projection of the Earth's equator into space, and the ecliptic is the Sun's apparent annual path around the sky.

Open the file "Intersection" to see how the vernal equinox is one of the two points where the celestial equator and the ecliptic intersect (the second point is called the autumnal equinox, and it is beneath the horizon). Start time running forward to see that the vernal equinox moves in the sky over the course of a night just like a star. It therefore has a fixed place in the celestial sphere, in the constellation Pisces.

Having defined the vernal equinox, the celestial meridian is then defined as the line extending from the North Celestial Pole through the vernal equinox to the South Celestial Pole.

Once a meridian has been defined, an object's right ascension is determined by its distance from this meridian. An object with an RA of 12 hours has the same RA as the autumnal equinox, while objects with an RA of 6h or 18h are halfway between the vernal and autumnal equinoxes.

Declination is perhaps easier to understand than right ascension, because its reference line is the celestial equator, which is just the projection of Earth's equator onto the celestial sphere. It is therefore an exact counterpart to latitude on Earth. We measure declination in degrees north or south of the celestial equator, which itself has 0° declination. The declination of the North Celestial Pole is +90° or 90° N and the South Celestial Pole is -90° or 90° S.

The equatorial coordinates of a star do not change as the hours pass (Actually, this is not precisely true because of **proper motion** and **precession**, but these motions are very slow, and do not change the position of a star enough to notice, at least with the unaided eye, during a human lifetime. See chapters 3.3 and 5.2 for more on proper motion and precession, respectively.) Just as the longitude and latitude of Boston do not change as the Earth turns, the equatorial coordinates of Rigel remain constant night after night and year after year. The equatorial coordinates of a planet or comet **do** change from night to night as the planet or comet moves across the sky relative to the stars beyond.

It is very useful to have a feeling for the equatorial coordinate system if you wish to be able to find things in the sky. An invaluable exercise is to stand outside at night and imagine the coordinates as if they were chalked on the sky, and to imagine how they turn with the celestial sphere as it

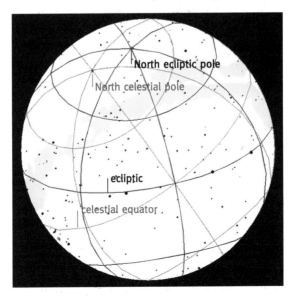

wheels overhead. With Starry Night, you don't have to imagine – just choose "Guides | Celestial Grid" and let time run forward. The equatorial coordinate system is based on the rotation of the Earth and is most useful for specifying the position of an object on the celestial sphere in relation to the Earth (to the Earth's equator and poles). A third system, called the *ecliptic coordinate system*, is more useful when you want to specify the position of an object relative to the Sun and to the orientation of the solar system.

The counterpart of the equator in the ecliptic coordinate system is the ecliptic, which is the Earth's path around the Sun. It is also the Sun's apparent annual path around the sky – another way of expressing the same concept. Positions in the ecliptic coordinate system are specified in *ecliptic longitude* and *ecliptic latitude*, both in degrees. The ecliptic itself has an ecliptic latitude of 0°, and the ecliptic poles have ecliptic latitudes of +90° and –90°. The North Ecliptic Pole – the north pole of the solar system – is in Draco, and the South Ecliptic Pole is in Dorado not far from the Large Magellanic Cloud. Positions along the ecliptic are measured eastward in degrees from an arbitrary starting point, which is again chosen to be the vernal equinox – the point where the ecliptic intersects the celestial equator in Pisces. Ecliptic longitudes are numbered between 0° and 360°.

Specifying an object's position in ecliptic coordinates tells you when the Sun is closest to it and how far it is from the Sun's path. The ecliptic position of Antares, for example, is 250°, -5°; the Sun is closest to longitude 250° on December 2 (you have to look up the date – or let Starry Night show you), and at that time Antares is 5° south of the Sun. Ecliptic coordinates are almost exclusively used for solar system objects. Planetary conjunctions are usually expressed in ecliptic coordinates (see the section on conjunctions in Chapter 4.2). When Apollo astronauts left the Earth's vicinity to travel to the Moon, they left the Earth's coordinate system behind and used the ecliptic system to navigate. Special ecliptic-based star charts were prepared for their voyages.

A fourth coordinate system is of specialized use to people interested in the Milky Way. It is the **galactic coordinate system**, and it uses the equator and poles of the Milky Way Galaxy as its fundamental plane and polar points. It lets you specify the position of an object in an "east-west longitude" sense relative to the center of the Milky Way and in a "north-south latitude" sense relative to the plane of the Milky Way. The Milky Way's center is in Sagittarius (at coordinates 17h 46m, -29° in the equatorial system) and its north and south poles are in Coma Berenices and Sculptor (at coordinates 12h 51m, +27° and 0h 51m, -28°) respectively. The system lets you specify a star's position as its **galactic longitude** and **galactic latitude**.

There are many other less well-known coordinate systems, such as those associated with other planets or moons. These coordinate systems are used by solar-system astronomers when calculating planetary spacecraft trajectories and other phenomena associated with particular planets.

The most important coordinate systems for finding your way around the sky are the Alt-Az (horizon) and equatorial coordinate systems. The coordinates you will find in a book are usually equatorial, but you need the horizon coordinates to actually know where in the sky you should be looking! Starry Night offers an easy way to get the different coordinates of any object. Just double-click on the object to bring up an Info Window. If you click just below the "Set Time" on the right side of this window, a drop-down menu that lists the four coordinate systems we have discussed (there are actually two different options for equatorial coordinates shown. The difference between these two is discussed in Chapter 3.3). When you select one of these options, the appropriate coordinates show up beneath this pop-up menu. This makes it easy to convert between different systems. Remember that horizon coordinates are time-dependent, so make sure to set the time in Starry Night to the time at which you want to observe. Once you have a good understanding of coordinate systems, you will spend less time figuring out where to look, leaving you more time for the observations themselves!

# Section 3

# Earth's Celestial Cycles

A S WE SAW in Section 1, there are several
sources for the apparent motions of the stars,
planets and other objects that we see in the night sky.
We can divide these motions into two categories: the
actual motion of the objects, and the apparent motion
caused in reality by the Earth's movements. This
section of *Starry Night* looks at the Earth's
movements. The apparent motion due to the Earth can
be subdivided into three distinct components: rotation,
revolution and precession. These sources of
motion are all periodic, meaning that their
effect is repeated after a certain
time. However, the periods for
the three motions are
radically different: the
Earth completes one
rotation in a day, one
revolution of the Sun
in a year, and one
precession cycle on its
axis in 26,000 years!
Rotation is covered in
Chapter 3.1, revolution
in Chapter 3.2, and
precession in Chapter 3.3.

# ROTATION

## What's in a Day?

The time it takes the Earth to spin once is a day, but how long is that? Twenty-four hours is the fast and simple answer, but it is just one of several that are correct. "Relative to what?" is the second half of the question.

Our lives are regulated by the appearance (and disappearance) of the Sun, and the *solar day* is the fundamental unit of time. The average time from one sunrise to the next is one solar day, and it does equal exactly 24 hours. That is the time it takes the Sun, on average, to return to the same position relative to **you** and your horizon ("on the average" is an important qualification because the length of the day changes with the seasons; see Chapter 3.2 for seasonal changes in the Sun's motion).

We could also think of the time it takes the Earth to spin once relative to the stars. Because the Sun moves eastward a slight distance relative to the stars each day, during a ***sidereal day*** ("sidereal" means "stellar") the Earth turns less than once relative to the Sun while it turns exactly once relative to the stars. This makes a sidereal day shorter than a solar day by four minutes. The sidereal day is 23 hours 56 minutes and 4 seconds long.

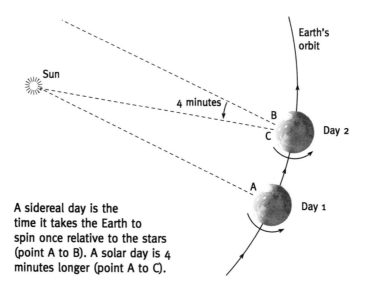

A sidereal day is the time it takes the Earth to spin once relative to the stars (point A to B). A solar day is 4 minutes longer (point A to C).

Open the file "Day" and notice that the Sun is on the meridian – the imaginary line that runs from due south to straight overhead and on to the north horizon. Daylight is turned off so that we can see the stars as well as the Sun. Some stars near the meridian are labelled.

Click the Single Step Forward button to advance the time by one sidereal day. The stars have not moved; more precisely, they have made one complete circle around the North Star and returned to their original position. But notice that the Sun is now about 1° to the east of its position the day before and is not yet on the meridian. Notice also that the date has advanced by one day but the time has decreased by 4 minutes. Now change the time step to 4 minutes and press the Single Step Forward button to complete one solar day. The Sun is now back in the same place in the sky, but the stars have shifted slightly to the west. This four-minute period is the difference between a solar and a sidereal day – the difference between one rotation of the Earth relative to the Sun and one rotation relative to the stars. Change the time step back to sidereal days and continue to step forward in time to make the concept clear: a sidereal day is one rotation of the Earth relative to the stars, not to the Sun.

## Hours, Minutes, and Seconds

We divide each day into 24 hours, each hour into 60 minutes, and each minute into 60 seconds. These are arbitrary divisions, and we could just as well divide the day into 10 hours of 100 minutes each, as was done following the French Revolution (the French decimal day was unpopular and did not last). Our hours come from ancient Egypt, where the day was divided into 10 "hours" of equal length with one more each for morning and evening twilight. The night was divided into 12 equal divisions for symmetry, giving a total of 24 hours for the complete cycle. The custom of dividing periods of time into units of 60 dates to the Sumerians, who liked to use numbers that were evenly divisible by many other numbers; 60 is evenly divisible by 30, 20, 15, 12, 6, 5, 4, 3, and 2.

# Time Zones

Early in the 19th century, most large American cities maintained an observatory, a major function of which was to determine the local time. The time in Detroit was different than the time in Toledo, and in fact it was different in every city across the land. The custom of using a unique local time in each community became more confusing as the speed of communication improved, and with the advent of rapid rail travel it became intolerable. It was impractical to maintain railroad timetables that had to allow for not only the speed of the train but also for local times at each station, and Canadian railroad magnate Sir Sanford Fleming lobbied successfully for *standard time zones*.

In 1884 the Earth's globe was divided into 24 meridians, each about 15° of longitude apart, starting with the zero (or prime) meridian, which passes through London, England. Each time zone is one hour wide. All areas within a time zone have the same solar time, and it is one hour later in the time zone to the east. The time zone boundary lines are not perfectly straight, but generally follow political and natural features for the convenience of people living within them. This means that it is not possible to determine your time zone from your longitude alone (as an extreme example, China spans about 60° of longitude, but the entire country shares the same time zone). Starry Night automatically adjusts the time zone if you select a city from the list in the Viewing Location window. If your city is not listed and you have to enter your viewing coordinates directly, you will also need to specify the time zone. You can find time zone information from many sites on the Internet.

Having standard time zones is a great convenience, but there are astronomical compromises. It may be the same moment by the clock at all locations within a time zone – but the Sun and stars are not at the same place in the sky for all those locations. If the Sun sets at 7:01 P.M. for a city at the eastern end of a time zone, it will still be well above the horizon as seen from a city near the western edge of the same time zone, where it will not set until almost 8:01 P.M. Likewise, moonrise and moonset times are not the same for all locations within a time zone. Starry Night gives you customized sunrise, sunset, moonrise, and moonset times for your location, which may differ from times published in local newspapers.

To see how rise and set times differ between localities, use Starry Night to find sunrise and sunset times from your home location. Remember that this information is contained in the Sun's Info Window, which you can show by double-clicking on the Sun. Now change your viewing location to another city in the same time zone (select a city by choosing "Go | Viewing Location" from the menu and pressing the Lookup button) and get the new rise and set times to see the difference. Notice that moving north or south has much less effect on the rise and set times than moving an equivalent distance east or west.

A further twist to time zones came with the introduction of Daylight Saving Time during World War I. Originally called "War Time," the purpose of moving sunset back one hour was to reduce fuel consumption for lighting and increase production in poorly lit factories during the late evening. It was dropped at the end of the war, and then reintroduced during World War II. It was dropped in 1945, and reintroduced most recently in 1967. It presently begins on the first Sunday in April and ends on the last Sunday in October in most of the United States and Canada; in Europe (where it is called Summer Time) it runs from the last Sunday in March until the last Sunday in September. Daylight time is great for kids who like to play outside late at night, but it delays stargazing by the same hour.

If you have your computer clock set correctly, Starry Night will detect whether or not daylight time is in effect for your home location on the current date. The small Sun icon shown to the left of the time tells you if daylight time is on; when the icon of the Sun is bright yellow, daylight time is on, and when the icon is dimmed, daylight time is off. Just click on this icon to turn daylight time on or off.

At any given moment – right now, for example – it is 24 different times around the Earth – one time for each time zone (a messy complication is that some jurisdictions have chosen to set their clocks using fractions of a time zone and are one-half or even one-quarter of an hour off standard time). This is a problem if you want to publish the time of an event in a table that is useful for the entire world. Astronomers find it effective to express events in *Universal Time*, abbreviated UT, which is the local time at the Greenwich Observatory in London, England. You need only know the difference between your clock time and Greenwich time to apply corrections (for example, Eastern Standard Time is five hours earlier than Greenwich). Converting local times back to Universal Time when communicating with other observers around the world is a convenience to people who then need only to apply the familiar conversion of Universal Time to their own time.

Astronomers often need to know the interval between two times separated by days, months, or years – for example, the exact interval between two observations of a variable star made months apart. Our calendar's many irregularities make it laborious to find, for example, the interval between 1:17 A.M. March 15, 1988 PST and 4:06 A.M. September 12, 1999 EDT. The *Julian Day* system lets you do this easily. This system assigns consecutive numbers to days starting at an arbitrary "day zero" in 4713 BC. To further simplify calculations, the Julian Day system uses decimal days rather than hours and minutes. Convert each of the two dates to a Julian Day, and then subtract. In the example above, the first date converts to Julian Day number 2447235.88681 and the second to Julian Day number 2451433.83750, and subtract; the answer is 4197.95069 days.

# The Nightly Rotation of the Sky

The sky turns overhead day and night, rotating as if it were a giant crystalline sphere (which it was thought to be during Medieval times) with the Sun, Moon, planets and stars attached to it. It rotates as a single unit at the rate of 1 r.p.d. (rotation per day).

As you saw in the exercise for Chapter 1.2, the North Star, Polaris, stays motionless in the sky. It is the pivot around which the rest of the stars turn. Or at least it appears to be. Open the file "NCP" to watch the rotation of the stars. Now zoom in using the Magnify button until your field of view is about 10°. You can now see that Polaris is not really motionless, and it makes a circle like any other star. The true pivot point is **the North Celestial Pole**, which is the point in the sky directly above Earth's north pole. Turn on the display of the North Celestial Pole by selecting "Guides | Celestial Poles." Polaris is not exactly at the sky's pole, but is 3/4° from it – pretty

## Revolution vs. Rotation

Two words that are confused endlessly are **rotation** and **revolution**. Objects rotate on their axes and revolve around another object. The Earth rotates once a day and revolves around the Sun once a year. The word rotate has an **a** for **a**xis, which is what an object rotates on.

close! This means that it makes a circle around the North Celestial Pole, 1-1/2° in diameter, and the North Star is never more than 3/4° from true North – too small a discrepancy to make a difference for most navigators. When you have finished watching the motion of Polaris, close the file "NCP."

Stars near the North Star make small circles around it, completing one circle in 24 hours. Stars farther from the North Star make larger circles, but still one circle per 24 hours. Stars far enough from the North Star set below the northwest horizon, disappear briefly below the northern horizon, and then rise again in the northeast.

**Long time-exposure photograph centered on the North Star.**

The far northern stars that do not rise or set at all are called circumpolar (in the Southern Hemisphere, it is the far southern stars that are circumpolar). They were considered immortal in ancient Egypt and they were associated with the pharaoh – who also was immortal, and who ascended to and lived among the far northern stars once his life on Earth was finished. Although we do not think of them as sacred today, the circumpolar stars have the distinction of not setting and thus are theoretically visible through the year (theoretically because they may come too close to the northern horizon to see in practice). The Little Dipper, Cassiopeia, Cepheus,

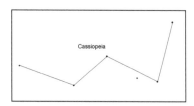

**The familiar W made by Cassiopeia's brightest stars is circumpolar to most observers in the Northern hemisphere.**

and most of Draco are circumpolar as seen from Canada and most of the United States and Europe.

How a star appears to move across the sky as the sky rotates depends on its position on the celestial sphere. If you live at mid-northern latitudes (anywhere within the United States and Canada below the Arctic Circle), a star (or planet, or the Sun, or Moon) that rises due east reaches its highest point when it is due south (when it is on the meridian), and it sets due west. Stars that rise south of due east do not reach as high a point on the meridian, and they set correspondingly south of due west. Stars that rise far to the south of east make short arcs across the southern sky, are not very high when on the meridian, and set shortly after rising in the southwest. You can easily visualize that there is a portion of the sky that lies farther south than the part you can see and that remains below the southern horizon.

Now we will see how a change in your latitude changes the rotation of the sky. We will begin at a latitude of 15° N by opening the setting "15N." Note how low Polaris is in the sky. Find the stars Vega and Eltanin on the left side of the screen. Now press the Forward button in the time controls to start time running. Both Vega and Eltanin fall beneath the horizon before reappearing.

Open the file "45N" to see how the sky looks from a latitude of 45° N. The North Star is farther above the horizon. Again find the stars Vega and Eltanin on the left side of the screen and start time running. At this latitude, Eltanin is circumpolar, while Vega still falls below the horizon briefly before reappearing.

Now open the file "60N" to view the sky from a latitude of 60°□NThe North Star is still higher. Again find Vega and Eltanin and start time running forward. From this latitude, you can see that both Vega and Eltanin stay above the horizon at all times. It is a rule that as you move farther from the equator, the number of stars that are circumpolar increases. At the same time, a proportionately greater area of the southern sky remains invisible below the southern horizon, and stars you might have once seen near your southern horizon no longer rise into view.

These ideas are carried to their logical extreme at the North Pole. Open the file "90N" to view the sky from the North Pole. From here, the entire sky is circumpolar. Start time running to see that no stars rise and no stars set; they all remain visible all the time, and each star maintains a constant altitude above the horizon. The North Star cannot be seen. This is because it is directly overhead. Remember that the altitude of the North Star is equal to your latitude (as you learned in the exercise for Chapter 1.1), which is now 90°. If you scroll up in the sky, you can see the North Pole almost exactly at the zenith.

Scroll around to the south to find Saturn and Jupiter. If you follow their motion, you can see that they (like all the planets, as well as the Moon and Sun) also circle endlessly at a constant elevation, neither rising or setting. In Chapter 4.2, we will see how and why these bodies do rise and set as seen from the pole; they do rise and set, but not because of the Earth's rotation.

The flip side of being at a location where all stars are circumpolar is that many stars and constellations that are prominent from mid-Northern latitudes cannot be seen at all from the North Pole. Choose "Edit | Find" and type "Rigel." A window will come up telling you that Rigel is beneath the horizon. You can run time forward and wait as long as you want, but this bright star in Orion will never come above the horizon.

From any location, half of the sky is visible at any given moment. However, the closer you are to the equator, the greater the number of stars that rise and set, meaning that you can see more different stars, but they are above the horizon for a shorter period of time.

Now move to the equator by opening the file "Equator." You can't see the North Star although it is directly in front of you, because it is right on the horizon. Its altitude is equal to your latitude, which is now 0. If you start time running forward, you can guess its location by seeing how the stars still pivot around a point. Now scroll around so that you are facing south. The stars in this half of the sky pivot around a point on the horizon that is due south. This is the *South Celestial Pole*. The way the sky turns in the north is symmetrical with the way in turns in the south. Stars that rise due east pass directly overhead and set due west. From this location, all stars rise and set; no stars are circumpolar. Along the equator (and nowhere else on Earth), every part of the celestial sphere is visible at some point in time.

Now we will move south. Open the file "30S" to view the sky from a latitude of 30° South. A large area of the sky that includes the Southern Cross is circumpolar. Notice that there is no bright South Star (see the section on Crux in Chapter 5.3) and the sky rotates around an empty, starless point (the few stars near the South Celestial Pole are quite faint, and can only barely be seen in perfect conditions, making them almost impossible to use as navigational aids). Scroll around so that you are facing north, and you will see familiar northern constellations in unfamiliar positions (unless, of course, you live in the Southern Hemisphere); some are upside down and others are on their side. The Sun still rises in the east (the Earth still rotates so that the sky turns from east to west) – but the Sun moves across the northern sky, not the southern as we see it from North America.

Finally, travel to the South Pole by opening the file "90S." The South Celestial Pole is directly overhead. All visible stars are circumpolar, and each star maintains a constant elevation as the sky turns. The sky rotates in a similar manner as from the North Pole, but the stars you see are completely different. Not a single star in this hemisphere of the sky can be seen from the North Pole! If you scroll around, you will see the Southern Cross two-thirds of the way up the sky, Sirius is 17°

**No bright star stays near the South Celestial Pole.**

above the northern horizon, and Orion is upside down with his head below the horizon and his feet in the air.

From everywhere on Earth, the sky rotates once a day, but how the stars rise and set, which are circumpolar (and which are visible at all), depends on your latitude.

# REVOLUTION

The second major motion of the Earth is its yearly revolution around the Sun. It affects how we see the Sun, stars, and planets.

## The Year

Just as the day is a unit of time based on the Earth's rotation on its axis, the year is based on the revolution of the Earth around the Sun. The time it takes the Earth to orbit the Sun once – or the Sun to circle around the sky relative to the stars – is one year. In Chapter 2.2 we found that there are several definitions for something so simple as the length of the day, and it is the same with the length of the year. The length of the annual cycle depends on which reference point is used.

If we measure the time it takes the Sun to circle the sky once and return to the same position relative to the distant stars, 365 days, 6 hours, 9 minutes, and 10 seconds will pass. This is the *sidereal year*, sidereal meaning "stellar" or "of the stars" in Latin.

However, it is not the most useful year for most purposes because it does not allow for precession.

Precession (which is described more fully in Chapter 3.3) is the slow wobbling of the Earth's axis in a 25,600-year cycle. Precession causes the vernal equinox – and all other points along the ecliptic – to shift in the same long 25,600-year cycle. If we used sidereal years for our calendars, the seasons would slip through the months and eventually the Northern Hemisphere winter would begin in July. As it is more convenient to keep the seasons constant – so that the vernal equinox always occurs on or near March 21, for example – we measure the year as the length of time it takes the Sun to circle the sky relative to the vernal equinox. The vernal equinox, you will recall, is the intersection of the celestial equator and the ecliptic. This point of intersection precesses slowly westward along the ecliptic, and it takes the Sun less time to return to the vernal equinox than to the same point relative to the stars. This so-called *tropical year*, on which our calendar is based, is 365 days 5 hours 48 minutes and 46 seconds long from vernal equinox to vernal equinox. The time it takes the Sun to circle the sky once and return to the westward-moving vernal equinox is a few minutes less than it takes the Sun to return to the same position among the stars.

Open the file "Vernal" and notice how the vernal equinox moves relative to the stars. Use the Single Step Forward button in the time controls to step forward through time and watch the vernal equinox slide to the right along the ecliptic relative to the stars. It is important to understand how slow this change is occurring. To notice it at all, you are zoomed in to a very high magnification and advancing time by steps of 365 days. Return to a regular 100° field of view and step forward a few more times. At this scale the movement of the vernal equinox with respect to the stars is imperceptible, which is why the difference in the lengths of the two types of years is only twenty minutes.

# Seasonal Changes In the Path of the Sun

You learned in the section in Chapter 2.2 that the ecliptic is inclined to the celestial equator. These two "great circles" (circles on a sphere) intersect at both the vernal equinox and the autumnal equinox. The angle these two great circles make with each other at that point is 23-1/2°, which is the amount the Earth is tilted on its axis. You can also think of it as the angular distance between the North Ecliptic Pole and the North Celestial Pole. The technical term for this tilt is the ***obliquity of the ecliptic***. If the Earth sat upright in its orbit (as Mercury does), the two poles would coincide – and so would the ecliptic and the celestial equator. This tilt is not just a curiosity – without it, we would not have seasons!

Unless you live on the equator, you know that the changing seasons bring changes in the daily motion of the Sun. The coming of winter means that the Sun rises later, sets earlier, and follows a lower path across the sky. When winter fades into spring and then

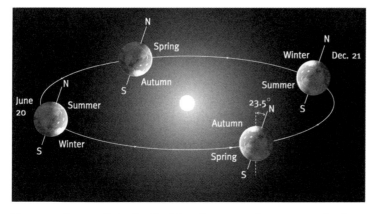

The 23.5 degree tilt in the Earth's axis creates the seasons. At the solstice on June 20, the Earth's Northern Hemisphere is at a maximum tilt toward the Sun, while the Southern Hemisphere is at its furthest tilt away from the Sun. The situation reverses at the other solstice on Dec. 21. The mid way point, called the equinox, defines spring and autumn.

into summer, these changes are reversed. The cause of these changes is the Sun's changing position along the ecliptic.

 Open the file "South" to watch the sky turn as a night passes. We are facing south with the celestial equator and ecliptic displayed, and have chosen to view from Anchorage, Alaska, because the celestial equator is relatively low in the sky and easy to see. Use the Single Step Forward button in the time controls to step forward through time by 5-minute intervals. Notice that the celestial equator (and all lines of declination) remains in the same position all night long. The celestial equator is an arc that extends from east to west and that has a maximum elevation, in the south, that is the complement of your latitude (90° minus your latitude). Points on the celestial equator move westward as the sky turns, but it is important to note that **the celestial equator itself remains in the same position all night long**. This is true no matter where you are on Earth.

In contrast, you can see that the ecliptic "wobbles" as the sky turns. In the Northern Hemisphere, the part of the ecliptic that runs through Scorpius and Sagittarius is low in the south when it is in view, while the part that runs through Taurus and Gemini is higher and much nearer to overhead when it is in view. This causes the Sun – which stays on the ecliptic – to be high in the sky in summer, when it is in Gemini, and low in winter, when it is in Sagittarius. If you keep stepping forward in time until the Sun comes into view, you will see that it is along one of the lower parts of the ecliptic, which is not surprising, because the month is December!

To see how the Sun's elevation changes during a year, open the file "SunElevation." The location is set for Iceland so the Sun will appear low in the sky, but after running this demonstration you should change to your home location and run the demo again. The Sun is locked and will stay centered in your screen.

We begin with the Sun on the meridian on June 21, which is the *summer solstice* – the first day of summer. Double-click on the Sun to bring up the Info Window. If you choose local coordinates, you will see that the Sun has an elevation (or altitude) of 49° as seen from Iceland. This is as high as it will be that day.

Now use the Single Step Forward button in the time controls to step through one year at the rate of one solar day per step, watching the Sun run eastward along the ecliptic while remaining due south and on the meridian. The Sun's elevation decreases daily, slowly at first – through July – but at a faster rate as summer ends. On August 1 it is 44° high at local noon. (You will see the Sun deviate from the meridian slightly. This deviation is called the *equation of time* and it is explained in the sidebar below) The rate at which the Sun loses altitude is greatest on the *autumnal equinox*, which is on or very near September 22. This should not surprise you, as you have probably noticed in your own experience that the length of the day (the amount of time that the Sun is above the horizon) shrinks very rapidly with the onset of autumn. September 22 is the date when the Sun crosses the celestial equator in a southward direction, and it marks the first day of autumn in the Northern Hemisphere. As seen from Iceland, the Sun has a meridian elevation of 26°. This is 23° less than its elevation on the summer solstice.

Following the autumnal equinox, the Sun continues to move lower at noon each day, but at a slower rate of change. Each day the Sun is a little lower at noon than the day before until, on December 21 or 22, the Sun reaches its lowest "maximum daily elevation." This is the *winter solstice*, which comes from Latin words for "sun-stand" because the Sun "stands still" (does not continue to lower altitudes) before reversing itself and increasing its elevation. From Iceland, this minimum noontime elevation on the solstice is a scant 2-1/2°! The Sun barely rises.

# Equation of Time

The Sun is not a uniform timekeeper for two reasons. First, the Earth's orbit is not circular, so the Earth's speed around the Sun is not constant. According to the laws of planetary motion discovered by Kepler almost four centuries ago, a planet travels fastest when it is closest to the Sun (because it feels a greater pull of the Sun's gravity when it is closer to the source of the pull). The Earth's orbit deviates from a perfect circle by less than 2%, but that is enough to notice. The Earth is closest to the Sun (at *perihelion*) early in January each year, and then it is moving slightly faster than when at its farthest point (at *aphelion*) early in July. The Earth's changing speed around the Sun makes the Sun appear to change speed as it moves on the ecliptic around the sky.

Second, the Sun travels along the ecliptic rather than the celestial equator. The ecliptic is inclined to the equator but we measure time along the equator. Even if the Sun traveled at a uniform rate along the ecliptic, its projected motion on the equator would change during the year.

The combination of these two effects means that the Sun is not in the same place in the sky at the same time each day. Its altitude (height above the horizon) and azimuth (direction along the horizon) both change over the course of a year. If we plot the Sun's position in the sky at noon each day over the course of a year, it  traces out a figure-8 path known as the analemma. To see this in Starry Night, select "Go | Analemma."

Clocks (and societies) run best at a uniform rate.

Your wristwatch ignores the seasonal variations in the speed of the Sun and displays time as if the Sun traveled at a constant speed. This imaginary uniform-speed Sun is called the "mean Sun" ("mean" meaning average). The difference between true (or apparent) solar time and mean solar time is called the **equation of time,** and it can be anywhere between -14 and +16 minutes. Your watch keeps mean solar time, but a sundial keeps true solar time.

After the winter solstice, the Sun reverses its downward trend and stands a little higher at noon each day. By the spring or **vernal equinox** on March 21, it is back to the same noontime elevation as on the autumn equinox: 48°. It reaches its highest noontime elevation at the summer solstice in June, when it "stands still" again before reversing course and heading southward. Follow the path of the Sun over an entire year.

The changing elevation of the Sun means that the relative lengths of day and night also change with the seasons. The winter solstice is the shortest day of the year in the Northern Hemisphere. Between the winter solstice and the summer solstice, the Sun's maximum elevation is increasing, so the Sun is above the horizon longer, hence the days become steadily longer. On the day midway between the two solstices - the vernal equinox- night and day are both 12 hours in length, a fact which holds true anywhere on Earth. After the summer solstice, the Sun's maximum elevation begins to decrease, so the days get shorter, and keep getting shorter until the winter solstice comes again, becoming the same length as the nights on the autumnal equinox. In the Southern Hemisphere, this pattern is exactly reversed.

# Effects of Latitude

The maximum daily elevation of the Sun changes with the seasons from every location on Earth, but the changes become more pronounced as you move farther from the equator. Two sets of latitude lines on the Earth are particularly important, and the Sun's behavior changes when you cross one of these lines. The two important sets of lines are:

1) the Tropic of Cancer and the Tropic of Capricorn
2) the Arctic and Antarctic Circles.

You will learn why these lines are important in the next two sections.

# The Tropics

In our view from Iceland, the Sun was never overhead. It reached a maximum noontime elevation of 49° on June 21, but went no higher. If you were south of Iceland, the Sun would be higher on the meridian on June 21 (and on every other date). If you were 5° south of Iceland, for example, the Sun would be 5° higher in your sky at noon on the same date. If you were just north of Mazatlán, Mexico, at a latitude of 23-1/2°, the Sun would just barely reach the zenith on the summer solstice. This latitude, called the ***Tropic of Cancer***, marks the northernmost limit on the Earth's surface where the Sun can be overhead. If you are north of the Tropic of Cancer, the Sun can never be overhead; if you are on the tropic, the Sun is overhead only on the summer solstice.

If you are south of the Tropic of Cancer, the Sun can be **north** of overhead. If you are a slight distance south of the tropic, the Sun is north of overhead at noon only a few days or weeks of the year. The farther south you are, the greater amount of time the Sun spends north of overhead.

## Sunshine and the Earth's Circumference

By the time of the ancient Greeks it was understood that the Earth is a sphere and that its sphericity causes the Sun to sit higher in the sky as seen from latitudes closer to the equator than from northern latitudes, where the Sun shines down at a shallower angle. In the third century BC Eratosthenes noted that the Sun shone vertically down a well in Aswan, Egypt, at the summer solstice, but on the same date it was 7° south of overhead at Alexandria, 800 kilometers (500 miles) to the north. He assumed that the distance to the Sun was so great that its rays struck all parts of the Earth on parallel paths, and correctly concluded that the distance from Alexandria to Aswan was 7/360 of the circumference of the Earth.

If you are on the equator, the Sun spends the same amount of time north of the zenith as south of it. Open the "Equator" setting and step forward a day at a time by using the Single Step Forward button to track the Sun's location at noon each day, as seen from the equator. When the Sun appears above the zenith on the screen, it is in the northern half of the sky, and when it appears beneath the zenith, it is in the southern half of the sky. The Sun is directly overhead at noon on the two equinoxes.

As you head south of the equator, the Sun spends more time north of overhead than south. It is exactly overhead, however, only twice a year.

The line of latitude that lies 23-1/2° **south** of the equator – the *Tropic of Capricorn* – marks the southernmost line on the Earth's surface where the Sun passes overhead. If you are south of this tropic, the Sun is always north of overhead.

The area of the Earth's surface between these two lines is called, simply, the "tropics." These names – Tropic of Cancer and Tropic of Capricorn – reflect ancient history. When the tropics were named, the Sun was in Cancer on the summer solstice in the Northern Hemisphere, and in Capricorn on the winter solstice. This is no longer the case, as you will learn in Chapter 3.3.

# Arctic and Antarctic Circles

The tropics mark the limits on the Earth's surface where the Sun can be overhead. Two other lines on the Earth's surface mark the extreme points where the Sun does not set – or does not rise. These are the *Arctic and Antarctic Circles*.

Move to the Tropic of Cancer by opening the file "WinterTropic." As with all points north of the Equator, the Sun is at its lowest noontime elevation on the winter solstice. Double-click on the Sun to bring up the Info Window, and read the Sun's elevation: 43°. This is 20° higher than the Sun's maximum elevation as seen from Chicago on the same date. From the tropics, the Sun is relatively high even at the winter solstice (which is why it is not cold in the tropics in winter). The farther north you are, the lower the Sun is at noon.

As we move northward along the Earth's surface, the maximum elevation of the Sun at midday on the winter solstice decreases. For every degree of latitude we move north, the maximum elevation of the Sun is one degree lower. At some latitude the Sun's elevation will be zero; on the winter solstice the Sun will just barely rise before setting immediately again. This line of latitude - called the *Arctic Circle* – lies 23-1/2° south of the North Pole at a latitude of 66-1/2°. If you move north of the Arctic Circle, the Sun does not rise at all on the winter solstice. The farther north you are of the Arctic Circle, the greater the number of days in the year when the Sun does not rise. At the pole itself, the Sun does not rise for six months!

Let's position ourselves at the Arctic Circle just before noon on the winter solstice by opening the file "ArcticWinter." Step through time at 10-minute intervals using the Single Step Forward button. The center of the Sun barely touches the southern horizon at noon before setting again.

Now open the file "ArcticSummer." You are viewing from the same location on June 21, the summer solstice. The Sun is much higher in the sky – it has a noon-time elevation of 47°. The Sun's rays do not warm the ground very effectively at such an angle, so it is safe to predict a cool June day for the Arctic Circle. Step forward through the rest of the day until sunset – and discover that there isn't one! The Sun barely touches the northern horizon at midnight, but it does not set. You are in the "land of the midnight Sun" – at least on this one day, the summer solstice.

If you have ever traveled far north – to Alaska, Scotland, or Norway – in the summer, you've noticed that it gets dark very late at night or not at all. It is amazing to be able to read a newspaper at 11 P.M. However, if you've been at the same place in mid-winter you've been equally amazed that it remains dark until late morning and is not terribly bright even at noon; people often have to use lights in the house all day long in winter.

The Arctic Circle is the southernmost latitude that experiences a midnight Sun, and then only for one day – the summer solstice. As you move north of the Arctic Circle, there are more days surrounding the solstice when the Sun does not set. At the North Pole, the Sun rises on March 21 (the vernal equinox) and sets on September 21 (the autumnal equinox) and there are six months of

## Daylight Quiz

Test your understanding. Which latitude has more hours of **total sunlight** during the year – the equator or pole? Think about it – you have enough information to figure it out.

Answer: None of the above. Ignoring clouds, all parts of the Earth's surface receive the same number of hours of sunlight during a year, which is a total of exactly six months worth. At the equator, the sunlight comes in 12-hour batches every day with little variance from one day to the next through the seasons; at the pole, it comes six months at a time, all at once, from March through September. Why, then, is it colder at the poles? Because the Sun's rays always strike the poles at a shallow angle, while at the equator the Sun shines down fiercely from nearly overhead 12 months of the year.

midnight Sun. That is followed by six months of no Sun (but not of no light – it takes a long time for the sky to get dark once the Sun has set at the North Pole.)

There is a corresponding latitude 23-1/2° north of the South Pole called the *Antarctic Circle* where all of this is true too, but six months out of phase. When the north polar regions are experiencing continuous sunlight, the south polar regions are in continuous darkness. When the Sun rises at the North Pole, it sets at the South Pole. When thinking about seasons and daylight, the Earth is very symmetrical about its vertical axis.

# The Sun Also Rises ... But Where?

So far we have been thinking about the Sun's height above the horizon at noon and how it changes during the year. You could monitor the Sun's height to track the seasons, and if you know the Sun's maximum altitude, you know the date. Let us now think about seasons in terms of the Sun's rising and setting points on the horizon.

Open the file "SunElevation2," and we are back in Iceland on the vernal equinox. It is 7:30 A.M. and the Sun has just risen. Notice where on the horizon the Sun rose; it rose due east, or 90° azimuth.

Step slowly through the calendar at one-day intervals for a few weeks, and stop. You will notice two things happening. Each day the Sun is **north of** and **higher than** its position the day before. It is higher each succeeding day at the same time (e.g., 7:30) because it rises earlier, and we have already discussed why this is so; now we want to keep track of the position on the horizon where it rises. Ignore the Sun's changing altitude, and resume running forward through time. The Sun's rising point moves northward through the months at a decelerating rate until it reaches its northernmost rising point; it pauses its northward travel before beginning to move south again. Note the date – it should be the summer solstice, June 21. Bring up the Info Window to see that the Sun rose at 3:02 A.M. Change the time to 3:02 A.M. to put the Sun on the horizon, and find its azimuth by bringing up the Info Window again. On June 21, as seen from Iceland, the Sun rises in the northeast at an azimuth of 21°.

The Info Window also shows that sunset will occur at 11:56 P.M. Change the time to this time and find the Sun again on screen, setting in the northwest. Bring up its Info Window again. If you check its azimuth, you will find that it is just under 339°, or 21° west of north (360 - 21 = 339). The Sun's setting position each day is always symmetrical with its rising position. If the Sun rises 30° north of due east one day, it sets 30° north of due west that same day.

At the summer solstice the Sun's rising point is as far north as it will ever get. On the following day, it begins to move south – slowly at first but at an increasing rate until the rate of change is greatest at the autumn equinox. It rises due east at the autumn equinox – as it does on the vernal equinox. For the next six months, the Sun rises in the southeast and sets in the southwest. Continue stepping forward in time to see this change. You may want to occasionally set the time ahead as you step through the days in order to keep the Sun near the horizon. Notice that by late November the Sun's rising point changes little from day to day, and the Sun again "stands still" on the winter solstice, when it reaches its southernmost rising point. From Iceland this point has an azimuth of 156°. This is as far south of due east (66°) as the Sun rose north of east on the summer solstice. As the winter solstice marks the Sun's lowest noontime elevation, so it marks the Sun's southernmost rising and setting points.

Perform the same experiment for your home town and find the Sun's northernmost and southernmost rising and setting points. Because the Sun rises at the same azimuth on the same date each year, if you know the Sun's rising point on the horizon, you know the date. (In practice, the Sun rises at the same azimuth twice during the year – except at the two solstices – but a clever person can keep the two straight.) Earlier cultures knew this too, and many kept track of the date by watching the point on the horizon where the Sun rose or set. In Russia early in the 20th century, one person in each village was appointed to sit in a certain place and monitor the sunsets to know when the festival days should be celebrated, and some native Americans still use the technique.

Finally, look at the Sun's changing and rising points from the Arctic Circle and then the North Pole and South Pole. Ask yourself what you expect to see before doing the experiments. By now, you should have a solid understanding of how the Earth's revolution around the Sun and its tilt combine to cause changes in the Sun's motion throughout the year as seen from all parts of the Earth. A tool like Starry Night is invaluable for such visualizations.

## Seasonal Changes in the Stars and Planets

While the changes in the motion of the Sun are the most dramatic effects of the Earth's revolution, our ability to observe the stars and planets is affected as well. This is due to the Sun's apparent motion through the celestial sphere along the ecliptic.

We saw in the exercise for Chapter 1.2 that constellations are sometimes classified as "winter" or "summer" constellations depending on the season when they are best observed. The general rule of thumb is that the farther a constellation is from the Sun, the longer it is visible at night. So when is the best time to see Scorpius? Not during the month of November, when the Sun passes through it, but six months later, when the Sun is on the other side of the sky. This rule of thumb works best for constellations near the celestial equator, for their distance from the Sun varies the most with the changing seasons. The farther a star is from the celestial equator, the more difficult it is to group it

**The Scorpion is a prominent figure during evenings in the late spring and summer.**

with a certain observing season.

The motion of the Moon and planets is more complex because we have to consider the apparent change in their position due to the Earth's revolution, as well as the true motion due to their orbit around the Sun (or the Earth, in the case of our Moon). The motion of the Moon is studied in Chapter 4.1, and the motion of the planets is considered in Chapter 4.2.

# PRECESSION

So far we've looked at two motions of the Earth which change the appearance of the sky: the Earth's daily rotation on its axis and its annual circuit around the Sun. The third motion of the Earth that changes the appearance of the sky, and that has been mentioned several times, is **precession**, also called "precession of the equinoxes." It is the wobbling of the Earth on its axis.

Precession was discovered in the second century BC by the Greek astronomer Hipparchus. He noticed a "new star," which we would call a nova or supernova (an exploding star), and wondered about the permanence of the stars. Although it was not customary for astronomers to make actual observations of the sky (such work was considered labor, and manual labor was assigned to slaves), Hipparchus decided to make an accurate record of the stars, their brightnesses, and positions. He compiled the first complete star chart – one that was used for centuries after. Like all star catalogs of the time, it used the ecliptic coordinate system. In the process, he compared stars he observed with observations recorded 150 years

## Worshipping the Secret of Precession

A mysterious religion, now extinct, once incorporated secret knowledge of the precession of the equinoxes. This knowledge was kept so private that only in the last few decades was it rediscovered. Yet at one time its followers were spread through the Roman Empire from England to Palestine and their religion was a rival to young Christianity.

Worshippers of Mithras portrayed their hero slaying a bull in the presence of figures of the zodiac. Taurus, the celestial bull, died in the sense that precession had moved the location of the vernal equinox from Taurus into Aries, ending the "Age of Taurus." A force that could move the equinox was stronger than any other yet known, for it moved the entire cosmos. Such a force must come from beyond the cosmos, and it was worshipped by the followers of Mithraism – who kept this knowledge a secret. They certainly would have been shocked to learn that the force originates with the pull of the Sun and Moon on the Earth's equatorial bulge!

previous by earlier astronomers, and he noticed a systematic shift in the ecliptic longitudes – but not latitudes – of the stars that was greater than could be accounted for by errors of measurement. He concluded that the coordinate system itself was shifting, although he could have no idea why.

Since the equinoxes – the intersections of the celestial equator and ecliptic – were shifting, and the vernal equinox marked the zero point of the ecliptic coordinate system, he called the motion the "precession of the equinoxes." His value for the annual precessional shift was within 10% of the correct value. In addition to discovering precession and compiling the first star catalog, Hipparchus devised the stellar magnitude system that is still used today, and was the first to specify positions on the Earth's surface by longitude and latitude. He is justly remembered as the greatest astronomer of antiquity.

Today we know that the Earth's axis wobbles and the equinoxes process largely because of the gravitational influence of the Moon. The Earth's relatively rapid 24-hour spin causes the Earth to bulge at its equator. The Earth's equatorial diameter is 21 kilometers (13 miles) greater than its polar diameter. The Moon – and to a lesser extent the Sun and to a far lesser extent the planets – pull on this slight bulge. The bulge is oriented along the Earth's equator, but the Moon and Sun pull from a different direction – from the ecliptic. The effect is to try to pull the Earth into a more upright orientation. But the Earth is spinning like a gyroscope, and it resists being pulled over. Instead, it precesses, or wobbles; the **amount** of tilt (in this case, 23-1/2°) remains constant while the **direction** of tilt changes. The Earth's axis traces a huge circle in the sky with a radius of 23-1/2° in a time span of 25,800 years (see illustration, right). That is a long time – there are as many human lifetimes in one wobble as there are days in one year. The celestial equator, which is always 90° from the poles, wobbles at the same time and at the same rate as the poles.

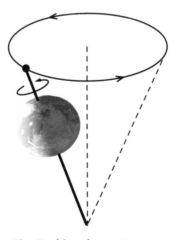

The Earth's axis rotates (precesses) just as a spinning top does. The period of precession is 25,800 years.

## Thuban and the Great Pyramid

The North Star at the time of the construction of the Great Pyramid was Thuban, an unassuming 4th-magnitude star in Draco the Dragon. You can find Thuban midway between Mizar in the Big Dipper and the end of the bowl of the Little Dipper. In 2700 BC the Earth's axis pointed near Thuban, and the star held special significance to the Egyptians, who associated it, and the "undying" stars that were circumpolar and that never set, with the Pharaoh. The northernmost "air shaft" leading upward from the King's Chamber in the Great Pyramid pointed to Thuban, symbolically connecting the dead pharaoh with the central undying star.

Thuban is corrupted Arabic for "serpent's head." Open the file "Thuban" to see how far Thuban is from the North Pole today.

The Earth's axis meets the sky at the North Celestial Pole (NCP), and as the axis precesses, the NCP changes. Right now, the NCP is less than 1° from the North Star, Polaris, and a century from now the NCP will be slightly closer to Polaris. Then the NCP will move on.

Five centuries ago, when Columbus sailed the ocean blue, Polaris was 3-1/2° from the NCP, and sailors had to make allowances for this offset when navigating. At the time of the birth of Christ, Polaris was 12° from the NCP and not a useful pole star at all. Almost thirty centuries earlier, Thuban – a faint star in Draco – was the pole star. Through the millennia, the Earth has had several pole stars – but most of the time there has been no bright star near the NCP.

At present, the North Celestial Pole is less than one degree from Polaris. It continues to move closer, and will be closest to Polaris (about 1/2 degree apart) around the year 2100 before moving away again. There will not be another pole star until about 4,000 AD when Er Rai (Gamma Cephei) will be fairly close. In 7,600 AD the star Alderamin (Alpha Cephei) will be the north star.

The three best north stars of the next 10,000 years. This view is from the year 4100 AD.

Our present pole star is the best the Earth will ever have. No other star is ever as bright and as close to the celestial pole as ours will be for the next two hundred years. It's one of those little things we just take for granted.

The sky's **pole** shifts against the stars due to precession, but so does **every** point on the sky. The entire equatorial coordinate grid precesses with respect to the stars, as Hipparchus discovered. The RA and Dec of a star are valid numbers only for a particular moment, called the *epoch* and the coordinates change with time. For example, the coordinates of Vega in 2000 AD are RA 18h 37m, Dec +38° 47', but in 2500 AD they will be RA 18h 54m, Dec +39° 22'. We must specify the epoch of the coordinates if we are dealing with long spans of time or extremely precise coordinates. Astronomers often use J2000 coordinates, which are the coordinates an object had in January 2000. Now you can explain why two different equatorial coordinate systems show up in an object's Info Window in Starry Night. One system gives the coordinates for the current date ("current" meaning whatever date you are looking at with Starry

Night), the other gives the coordinates for January 2000. If you compare the two systems, the coordinates will be almost identical. Precession only has a major effect over a very long period of time.

# Precession & Astrology

Precession has an interesting effect on astrology, and especially on *birth signs* or *astrological signs*. The signs – such as "Scorpio" – are each a uniform section of the sky 30° wide. They are measured eastward along the ecliptic from the vernal equinox, which is the intersection of the ecliptic and the celestial equator and is the zero point. When this system was set up around 600 BC, the zero point was in Aries and was called the "first point of Aries." The constellation Aries encompassed the first 30° of the ecliptic; from 30° to 60° was Taurus; from 60° to 90° was Gemini, and so on. This scheme ignored actual stars, but uniformity was more important than fussing about star positions.

Since then, precession has caused the celestial equator to wobble so as to cause the intersection point between it and the ecliptic to move westward along the ecliptic by 36°, or almost exactly 1/10 of the way around. The vernal equinox has also moved. It crossed from Aries into Pisces around the time of the birth of Christ (and the birth of Hipparchus), according to modern constellation boundaries. In six more centuries it will leave Pisces.

Your birth sign in your morning newspaper horoscope ignores precession. What your horoscope calls "Aries" is the 30° segment along the ecliptic that is east of the current location of the vernal equinox – but most of it is in Pisces! The next 30° segment, called "Taurus" in horoscopes, is largely in Aries. In the last 2,600 years the signs have slipped 2,600/25,800 of the way around the sky in a westward direction, relative to the stars beyond. The astrological signs are directions in space that do not correspond to the stars in the astronomical constellation with the same name.

Precession causes the position of the Sun on the vernal equinox to shift as the Earth wobbles – but so does the position of the Sun on **every** date. This means that it is not only the names of the Zodiac signs that are now inaccurate. The names of the **tropics** are now inaccurate as well.

Open the setting "Tropic," which shows the position of the Sun as seen from a viewing location along the Tropic of Cancer at the time of the summer solstice in the Northern Hemisphere, June 21. If you double-click on the Sun to bring up its Info Window, you will see that the Sun has an altitude of almost 90°, meaning it is directly overhead. But the Sun is not in the constellation Cancer! Use the Magnify button to zoom in and see that it is actually just inside the boundary of Taurus (it was in Gemini until only a few years ago). Similarly, on the Southern Hemisphere summer solstice on December 21, the Sun is in Sagittarius, not Capricornus. Why is there a discrepancy? The tropics were named around 150 BC, and the Sun was in Cancer on the summer solstice then. Likewise, it was in Capricornus on the winter solstice (summer solstice in the Southern Hemisphere). Since 150 BC, precession has shifted the summer solstice westward from Cancer to Gemini to Taurus and the winter solstice from Capricornus to Sagittarius. Therefore, the Tropic of Cancer should be renamed the Tropic of

The winter solstice has moved far from Capricornus in the 2,600 years since the Tropic of Capricorn got its name.

Taurus, and the Tropic of Capricorn should now be named the Tropic of Sagittarius. But these names too will be ephemeral, and the tropics will continue to shift. In another 23,200 years the cycle will have been completed and the Sun will once again appear in Cancer and Capricornus on the solstices.

This table lists the dates when the Sun is actually within the astronomical constellations of the zodiac, according to modern constellation boundaries and corrected for precession (these dates vary by up to one day from year to year). You will probably find that you have a new birth sign. If you were born between November 29 and December 17, you are an Ophiuchus!

| Constellation | Dates |
|---|---|
| Capricornus | January 20 to February 16 |
| Aquarius | February 16 to March 11 |
| Pisces | March 11 to April 18 |
| Aries | April 18 to May 13 |
| Taurus | May 13 to June 21 |
| Gemini | June 21 to July 20 |
| Cancer | July 20 to August 10 |
| Leo | August 10 to September 16 |
| Virgo | September 16 to October 30 |
| Libra | October 30 to November 23 |
| Scorpius | November 23 to November 29 |
| Ophiuchus | November 29 to December 17 |
| Sagittarius | December 17 to January 20 |

To confirm that these dates are correct, open the file "RealZodiac." Step forward through a year and notice the dates on which the Sun moves into a new constellation.

## The Dawning of the Age of Aquarius

According the popular '6os song, "it is the dawning of the Age of Aquarius." What does this mean, and when does that happy day arrive?

The name of an astrological "age" comes from the constellation the vernal equinox is in. During classical Greek times, the vernal equinox was in Aries and it was the "Age of Aries." By about the time of the birth of Christ, the equinox had precessed westward until it stood in Pisces, and the last 2,000 years has been the "Age of Pisces." When the equinox moves into Aquarius, the "Age of Aquarius" will begin. Using modern constellation boundaries, the equinox has 9° farther to precess before it enters Aquarius, and that won't happen until the year 2597. Apparently we have a bit of a wait before the dawning of the age of universal peace and love.

# The Extended Constellations of the Zodiac

Using modern constellation boundaries, the Sun travels through the traditional 12 constellations of the zodiac plus Ophiuchus. The Moon and planets orbit on paths that are inclined to the ecliptic, and they travel through additional constellations. Not counting Pluto, whose orbit is inclined a whopping 17° and which would add several more constellations to the list, these are the 21 astronomical constellations that are visited by at least one of the solar system's major bodies.

| Aquarius | Corvus | Libra | Sagittarius | Virgo |
| Aries | Crater | Ophiuchus | Scorpius | |
| Cancer | Gemini | Orion | Scutum | |
| Capricornus | Hydra | Pegasus | Sextans | |
| Cetus | Leo | Pisces | Taurus | |

Open the file "ExtendedZodiac" and step through time to follow the motion of the Moon. You can see that it oscillates around the line of the ecliptic. Imagine making a band which is twice the width of the maximum distance from the Moon to the line of the ecliptic. If you wrap this band around the ecliptic line, every constellation partially or fully covered by this band is touched at some point by the Moon. The naked eye planets that travel farthest from the ecliptic are Venus and Mars, so they make the biggest "bands" and travel through the most constellations.

# Section 4

# Our Solar System

PEOPLE who live in brightly lit cities cannot experience the awesome beauty of a dark night sky with its "millions" of stars and the Milky Way unless they travel away from home – but they can see the Moon and planets. Tracking the motions of the Moon and planets is perhaps the oldest astronomical activity and it is one that city folk can still fully participate in. It is a good way to remain connected to the sky. Starry Night excels at showing you how the Moon and planets move and where they will be in the future, and it lets you plan your skywatching and understand what is going on. Chapter 4.1 will look at the motions of the Moon, while Chapter 4.2 will cover the motions of the planets. It is important to remember that the apparent motion of a moon or planet as seen from Earth is a combination of two things:

1) the motions of the Earth (as discussed in the last three chapters)

2) the real motion of the moon or planet around its parent body

# THE MOON

As long as people have been looking up, they have been fascinated by both the motions and the changing appearance of the Moon. Scholars debate how much was known, when, and to what use monuments such as Stonehenge and others were put, but all agree that the Moon was monitored carefully and even worshipped from remote prehistoric times. We don't worship it today, but we can still follow its motions.

The Moon revolves about the Earth approximately 12 times per year or approximately once every 30 days. This relatively rapid rate makes the Moon's motion among the stars easily visible from night to night.

Because the solar system is relatively flat, the Moon and planets stay near the Sun's path along the ecliptic. The Moon's orbit is tilted 5° to the ecliptic, and so the Moon remains within 5° of the ecliptic. Half the month it is north of the ecliptic and half the month south, and it crosses the ecliptic at two points called the ***nodes***.

Of course, Starry Night will show you precisely where the Moon

## The Oldest Moon Observation

A series of notches cut into a reindeer bone found buried in a cave in France may be a symbolic tally of the nightly appearances of the Moon over a two-month span. This bone is 30,000 years old, suggesting that as early as the Ice Age people kept track of the position of the Moon and counted the passing of time in months.

is tonight or any other night (or day), and not only from your backyard but from any place on Earth. It is in almost the same position in the sky for everyone on Earth at the same moment, but not exactly. Observers separated by a large distance will see the Moon in a slightly different position relative to the background stars. This is an example of *parallax* – the apparent shift of a comparatively nearby object seen against distant objects when viewed from different positions.

Demonstrate by using Starry Night to view the Moon from two different cities at the same moment. Open the file "MoonParNY" file, and notice Saturn just above the Moon. This view is from New York City. Leave this window open and now open "MoonParDetroit," which shows the Moon and Saturn as seen at the same moment from Detroit. Align the two Starry Night windows so that you can see both views at the same time. Notice how Saturn – and the background stars – are in slightly different positions relative to the Moon, depending on which city you are viewing from. In reality, it is the Moon whose position has shifted relative to the far more distant planet and stars. Moving to the far side of the Earth can cause the Moon to shift its position by an amount greater than its own diameter! This is especially important in a solar eclipse or the occultation of a planet or star by the Moon – the Moon may align with the Sun or occult a planet as seen from one city, but not from another.

# New & Full Moon

The Moon's monthly orbital cycle begins at ***new Moon***, when the Moon is in line with the Sun. In reality the Moon is usually a bit above or below the Sun then and not directly in front of it, so new Moon is defined as the moment when the Sun and the Moon have the same longitude, which is also approximately when they have the smallest angular separation. After one week, the Moon has moved a quarter of the way around the Earth and is at its first-quarter position. One week later finds the Moon opposite the Sun (or at least at longitude of the Sun plus 180°), and this is the ***full Moon***. One week later the Moon is three-quarters of the way around its orbit, and we call it a third-quarter Moon or last-quarter Moon. After another week it has completed its cycle and is new again.

## How Long is the Moon Full?

The Moon is full when it is opposite the Sun (when its longitude is 180° greater than the Sun's). It is full for only an instant, just as it is midnight (or 3:00 P.M.) for only an instant. That instant can be calculated to the nearest second, and it is the same moment simultaneously all over the Earth. To most people, however, the Moon looks full for three days to a week. For practical purposes, the Moon is considered to be full for the entire night closest to the instant of its opposition.

# The Month

The time it takes the Moon to complete one orbital cycle is a *month*, but how long is that? By now, you are probably not surprised to learn that there are two common definitions of the month. Open the file "MoonMonth" to find out why.

Begin with the new Moon on April 12, 2002. The time is 9:22 A.M. – the actual moment when the Moon is new – and we are viewing from Honolulu, Hawaii. The Moon and Sun are very close together. The stars are turned on and daylight colors are turned off. Notice that the Moon is just a fraction of a degree to the right of the star HIP 6751. Now step forward through time at one-day intervals until the Moon approaches HIP 6751 again. This will occur on May 9, which is a period of 27 days. The precise orbital period is 27.32166 days – the time the Moon takes to circle the Earth once relative to the stars, and it is called a *sidereal month*.

But notice that the Moon is not yet new! It is a thin waning crescent. During the 27-1/3 days it took the Moon to return to the same part of the sky, the Sun has moved on 27° eastward. Recall that the Sun moves eastward along the ecliptic because the Earth moves around the Sun, and that the Sun moves eastward with respect to the stars at the rate of very nearly 1° per day. In order to complete its cycle of phases and become a new Moon again, the Moon has to move that additional 27-1/3°, and that takes it 2-1/6 days. Step forward in time to see that the Moon and the Sun come closest together (and the Moon is completely dark) on May 12. This cycle of the Moon's phases – from new Moon to new Moon – has a period of 29.53059 days and it is called a *synodic month*. It is what we normally think of as a month.

The month is a fundamental cycle in the sky, and since prehistoric times it has formed the basis of a calendar. Early calendars were **lunar** – they were based on actual observations, and a new month began with the sighting of the new Moon. This is still true in Islamic culture. A problem is that there are not a whole number of lunar months in a year – there are 12-1/3 synodic months in a year. A calendar based on synodic months has to add leap months periodically if it is to keep the months in phase with the year. The Islamic calendar does not have leap months, so its months (and holy days) do not always fall in the same season. Our modern Gregorian calendar is solar, and months are given arbitrary lengths so they add up to 12 months of 365 days total.

The Moon's cycle also forms the basis of our week, which is one quarter of a synodic month (rounded off to the nearest whole day). The idea of the 7-day week originated in Babylon and was spread around the Mediterranean world by the Jews after they were released from Babylonian captivity.

## Motion of the Moon

Because the Moon stays near the Sun's path, if you know the movements of the Sun you already know the approximate movements of the Moon. There are differences, however. The new Moon rises and sets with the Sun and follows its path across the sky, but at its other phases the Moon travels a different path than the Sun. Let's look at a new and then a full Moon to see why this is so. Open the file "MoonPath"; the time is 7:00 A.M. on October 27, 2000, and the Sun and new Moon are close together and just above the southeastern horizon. Daylight is turned off so we can actually see both of them. Notice where the Sun and Moon are on the horizon as they rise. Step through time in 10-minute intervals and watch them move together across the sky until they set. Note the direction in which they set.

Now advance time by two weeks to setting the date and time to 5:30 P.M. on November 11. The Moon is now full, and it is again just above the eastern horizon. Notice how far north it is of its rising position two weeks earlier. Advance through time at 10-minute intervals and watch it move across the sky, and confirm that it sets far to the north of its setting position two weeks earlier. The full Moon is 180° opposite the Sun and does not follow the Sun's path for that day – it follows the path the Sun will have in six months, when the Sun is 180° in advance of its present position. The path of the full Moon is six months out of phase with the path of the Sun; the full Moon in winter travels the path of the Sun in summer, and vice versa. The Sun is highest in the sky (in the Northern Hemisphere) at the summer solstice in June, but the full Moon in June travels on its lowest path for the year. The full moons in December and January pass the highest overhead at midnight. Close the file "MoonPath."

## Midnight Moon

If you are north of the Arctic Circle, the Sun can remain above the horizon for more than 24 hours and you can see a "midnight Sun" (see Chapter 3.2). The Moon too can remain above the horizon for more than 24 hours in a row, and at other times it does not rise at all!

Open the file "MoonArctic1" to view the "midnight Moon" from northern Greenland. Step forward through time in 1-hour steps to watch it circle the horizon; at what time will it be due north? When will it finally set? Because the Moon traces the Sun's yearly path in only a month, the amount of time it spends above the horizon varies much more rapidly than does that of the Sun. Close the file "MoonArctic1."

# Surface of the Moon

Through even a small telescope you will see more features on the Moon than on the Sun, all the planets, and all the comets and deep space objects put together. It is beyond the scope of this book to interpret the Moon's features (consult a popular observing guide to the sky for this information), but let it be said that the Moon's surface is the result of impacts from above and lava flow from below. During its first half-billion years, the Moon was struck again and again by asteroids – debris left over from the formation of the planets. They exploded on impact as their kinetic energy was changed into heat, and they vaporized themselves and a great deal of the rock that they hit, blasting out huge craters. By four billion years ago the Moon's surface was so saturated with craters that not one square centimeter hadn't been hit at least once. Over the next billion years, immense amounts of hot lava flooded the low areas with smooth lava flows that cooled and turned dark. These colossal lava flows formed what we call the lunar "seas" – seas of cooled lava. On Earth, erosion from wind and rain has erased the evidence of the Earth's violent early history; but the Moon has no atmosphere and therefore no erosion. The lava stopped flowing long ago and few asteroids fall today, so the Moon looks much like it did three billion years ago.

It is a **very** common misconception that the Moon has a fixed dark side and a fixed light side. The Sun rises and sets on the Moon's surface as it does on the Earth, but with a "day" that is 27-1/3 Earth days long. Note however that the Moon **does** keep one side permanently

Sunrise as seen from the Moon, with a "new Earth" just to the Sun's right.

# The Earth from the Moon

Everyone has seen the Moon from the surface of the Earth, but only 12 men – the Apollo astronauts – have seen the Earth from the surface of the Moon. You can join them with Starry Night.

Open the file "MoonApollo11." You are at the Apollo 11 landing site along the Moon's equator. The Earth is high above the western horizon, in the constellation Pisces. Notice that the Earth is not full, but is partially in shadow. Double-click on the Earth to bring up an Info Window to discover its phase.

You cannot see the Moon half the time from your home on Earth, but there is no need to wonder if you will see the Earth from the Moon. If you are on the side of the Moon that faces the Earth, the Earth is always in the sky, day and night, and it never rises nor sets. It changes position little from day to day and year to year, but it does change its phase. The Earth goes through a full set of phases once a month just as the Moon does, but the two have opposite phases and remain 180° out of synchronization. When the Moon is new, the Earth is full, and vice versa. Think about it a minute, and then step forward through time with Starry Night. You will see the Earth rotate and the stars and constellations move behind it, but the Earth hardly changes its position in the sky.

The Earth is never visible from the far side of the Moon – the side that permanently faces away from Earth. Just as this side cannot be seen from Earth, visitors to it will never see the Earth in their sky.

turned toward the Earth while the back side (or far side) of the Moon is permanently turned away from Earth (i.e., its rotation and revolution periods are the same). The Moon is in what is called *locked rotation* with the Earth (as are several other moons in our solar system), and it is a long-term result of tidal forces between the Earth and Moon. All the familiar seas and craters are on the Moon's near side; the far side was unknown until photographed by a Soviet spacecraft in 1959.

# Eclipses

In the first section of this book we saw how Starry Night was able to simulate a solar eclipse. Both solar and lunar eclipses are caused by the motion of the Moon around the Earth. When the Moon moves directly in front of the Sun, there is a solar eclipse. When it passes exactly opposite the Sun in the sky, there is a lunar eclipse.

Eclipses are among the most awesome spectacles in nature, and one of the first duties of astronomers thousands of years ago was to predict eclipses – or at least to be on hand to perform the rituals that would cause the dreaded eclipse to hurry up and end! Only since about 600 BC could astronomers predict eclipses with any accuracy. With Starry Night, you can view any of the eclipses between 2000 and 2009. Select "Sky | Interesting Events" for a list of all these eclipses, and a description of whether the eclipse was solar or lunar, and whether it was total, annular or partial. Choose "Best View" to view the eclipse from the prime location or "Local View" to see what it looked like at that time for your home location.

If you do search for eclipses, you will notice that they do not happen randomly in time. The Moon's path is inclined 5° relative to the Sun's path, and an eclipse can happen only when the Sun is at or near a *node* of the Moon's orbit – one of the two places where the Moon's orbit crosses the ecliptic. When the Sun passes a node, at least one solar eclipse **must** happen, and there can be two. The Sun spends 38 days close enough to the node that the Moon can pass in front of it. If the Moon eclipses the Sun during the first 8 days of that period, the eclipse will be partial and the Moon will return and eclipse the Sun again one lunar month later,

Eclipses can only occur when the Sun is near the intersection of the ecliptic and the Moon's orbit.

yielding two partial eclipses. If the Moon does not eclipse the Sun until after the 8th day, there will be just one solar eclipse, and it will be total or annular, if seen from the best viewing location.

Open the file "EclipseNodes" to see the geometry of the eclipse nodes. Two lines are marked. One is the path of the Moon's orbit; the other is the ecliptic, the line which traces the Sun's apparent path. The intersection point of these two lines is an eclipse node, and it is just to the bottom left of the Sun. Step forward through time in one day intervals and you will notice that the Sun is moving closer to the node. At the same time, the Moon is rapidly moving around the sky, but it is far enough away from the Sun that it is off screen. On July 19, the Moon comes into view. At this time, the Sun is almost exactly at the eclipse node, so you can predict that there will be a solar eclipse when the Moon passes by the Sun. Continue stepping forward in time until July 22, when the Moon and Sun line up. As predicted, there is an eclipse. It is a total eclipse if viewed from this location on Earth, which is Shanghai, China.

Thousands of years ago astronomers realized that a solar eclipse could happen only during the 38-day period when the Sun was near

## Eclipses in Ancient History

Ancient eclipses are important to historians because they allow us to date events with precision. The dates of many important events are poorly known, often only to the nearest decade, but the times of all solar and lunar eclipses can be calculated to the nearest minute. If an eclipse accompanied an historic event, the time of the event can be pinned down precisely. Eclipses provide crucial "anchor points" in history.

one of the Moon's nodes, and this interval was called the *eclipse season*. Before astronomers were capable of predicting eclipses reliably, they issued warnings that an eclipse could happen during the eclipse season – just as today hurricane warnings are issued in the Caribbean during the storm season. The predicted eclipse did always occur, but it was often visible only from a distant part of the Earth and they were ignorant of it (they counted it as a miss). Today the only warnings astronomers issue regarding eclipses is to not look at the Sun without proper equipment.

As with the solar eclipse, both the Sun and the Moon must be near an eclipse node for a lunar eclipse to occur. During a solar eclipse, the Sun and the Moon are at the same node, but during a lunar eclipse, they are at opposite nodes, which are 180° apart in the sky. Open the file "EclipseNodes2" to see this. You are looking at the full Moon, which is dark and in shadow because we are witnessing a lunar eclipse. Double-click on the Moon to open its Info Window and read its azimuth. Now scroll around in the sky to find the Sun. You will see that it is also at a node. If you open its

Info Window, you will see that its azimuth is about 180° removed from the Moon's. There must be at least one lunar eclipse each eclipse season and it can be total, or there can be two – but then both are partial.

Eclipse season is the same for both solar and lunar eclipses. After a solar eclipse, the Moon must move about 180° in the sky for a lunar eclipse to occur. The Moon makes a complete 360° circle in about four weeks, so it moves about 180° in two weeks. We would therefore expect lunar and solar eclipses to be separated by about two weeks and indeed, this is the case. Every eclipse season brings at least one solar eclipse and at least one lunar eclipse.

If the Moon crossed the Sun's path at the same point each year, eclipses would happen on the same date each year, but it doesn't, so they don't. The Moon's orbit regresses in a motion that is very similar to the precession of the equinoxes (see Chapter 3.3). The Moon's

## The Last Total Solar Eclipse

The Earth has enjoyed total solar eclipses since the Moon was formed, but our distant descendants will not see any. Because of tidal friction with the Earth, the Moon is receding from us at the stately rate of 4 centimeters (1.5 inches) per year – about the same rate that the continents drift or that your fingernails grow. The Moon's apparent diameter is shrinking as its distance increases. In about the year 600,000,000 AD the Moon will totally eclipse the Sun for the last time; after this time the Moon will be so distant that its disk will no longer be large enough to completely cover the Sun, and people will see only annular and partial eclipses.

two nodes precess westward along the ecliptic at the rate of 18.6° per year. The Sun moves 18.6° along the ecliptic in 18.6 days and arrives at the node earlier the next year, causing "eclipse season" to move backward through the calendar. The eclipse of December 14, 2001 is followed by an eclipse on December 4, 2002.

# Viewing a Lunar Eclipse

A lunar eclipse is a direct (although less visually spectacular) counterpart of a solar eclipse. The Sun is eclipsed when the Moon moves in front of it and we find ourselves in the Moon's shadow. The Moon is eclipsed when it moves into the Earth's shadow. This happens only at full Moon – when the Moon is directly opposite the Sun.

The Earth's shadow – like all shadows – has a central ***umbra*** and an outer ***penumbra***. The umbra is the place where, if you stand there, the Sun is totally blocked and the eclipse is total. The penumbra is the place where the eclipse is partial. A lunar eclipse has stages that resemble a solar eclipse, and tables of eclipse circumstances will list the times of the several "contacts" (when the Moon begins to enter the penumbra, is fully within the penumbra, enters the umbra, is fully within the umbra, mid-eclipse, begins to leave the umbra, begins to leave the penumbra, and is finally out of the penumbra). Starry Night will show these several stages graphically and you can determine the timings from the date/time display. Open the file "LunarTotal" and step through time to view a total lunar eclipse. Note the faint penumbra and the darker umbral shadow that the Moon passes through.

During a shallow penumbral eclipse, when the Moon passes through only the outer edge of the Earth's shadow, the Moon may not darken noticeably. In a deep penumbral eclipse the darkening is barely noticeable even to a careful observer. In an umbral eclipse, one portion of the Moon grows dark and a casual observer would notice that something is amiss. Only during a total eclipse does the Moon darken substantially and take on a reddish or orange color. This coloring comes from light refracted around the edge of the Earth towards the Moon; it comes from all the Earth's sunrises and sunsets. If the Earth's atmosphere is especially opaque, as happens following a major volcanic eruption, the eclipse can be so dark that the Moon disappears, but this is rare. Generally the Moon takes on a deep coppery color, dims as the stars come out, and looks very pretty. To superstitious people in former times, the red color of the Moon made it an unpleasant and fearsome omen.

The Earth casts a much larger shadow on the Moon than the Moon does on the Earth, because of the Earth's larger size. This means that many more people will see a total lunar eclipse than a total solar eclipse for three reasons:

1) total lunar eclipses are slightly more common than total solar eclipses (there are about 3 total lunar eclipses for every 2 total solar eclipses) because the Earth, Moon, and Sun do not have to be **exactly** in a straight line for a total lunar eclipse to occur.

## The Greek's Proof of a Round Earth

The ancient Greeks knew that the Earth is a sphere by observing lunar eclipses. They saw that the edge of the Earth's shadow on the Moon always has a circular shape, and they knew that the only object that only casts a circular shadow is a sphere. They did not, however, know the size of the Earth or Moon or the distance to the Moon.

2) most total lunar eclipses are total as seen from almost anywhere on Earth where the Moon is above the horizon at the time of the eclipse, while total solar eclipses are only total as seen from a narrow path.

3) total lunar eclipses generally last much longer than total solar eclipses.

Unlike a solar eclipse, you need take no special precautions to observe a lunar eclipse. The Moon is a dark rock sitting in space, and when eclipsed it just grows darker still. Use binoculars or a telescope to bring out subtle coloring.

# Occultations

Another interesting event to observe with binoculars or a telescope is the *occultation* of a planet or bright star by the Moon. As the Moon circles the Earth, it occasionally passes in front of a more distant object and "eclipses" it. If the object eclipsed has a much smaller apparent size than the Moon (a planet or star, for example, but not the Sun), then the preferred astronomical term is "occultation," from the Latin root "occult," which has nothing to do with the supernatural but simply means "to conceal or hide."

A star, whose apparent diameter is miniscule, blinks out instantaneously when the Moon's edge moves in front of it, and it is startling how abruptly a star disappears from sight. Only stars within 5° of the ecliptic can be covered by the Moon, so we can only see star occultations for a handful of bright stars (Spica and Regulus are the most prominent).

A planet takes several seconds to disappear as the Moon moves in front of it. Up to an hour and a half later, the star or planet reappears from behind the other side of the Moon. Depending on the Moon's phase, the leading edge will be light while the trailing is dark (full Moon to new), or vice versa (new Moon to full). All planets eventually cross the ecliptic, so any planet can be involved in a planetary occultation.

Both types of occultations are shown in the files "Occultation1" and "Occultation2"; Open these and step forward through time slowly to watch Saturn and the star Spica respectively disappear and then reappear. The more readily visible occultations are listed in monthly sky calendars such as the one appearing in *Sky & Telescope* magazine, and at the International Occultation Timing Association's web page, www.occultations.org.

# PLANETS, ASTEROIDS & COMETS

## Planets

Chapter 1.3 of this book showed us a few fundamental aspects of the planets in our solar system: the Earth is a planet like the others in motion about the Sun; all of the planets (with Pluto the major exception) revolve around the Sun in the same plane; and the planets move in the same direction but at different speeds. Humans have required a significant fraction of their collective history to arrive at these ideas. It was especially difficult to discard the idea of the stationary Earth.

In the exercise for Chapter 1.3, we viewed the motion of the planets with the Sun at the center, which we now know is the correct view. But how would the planetary motions look if the Earth was really stationary? To find out, click on the "Go" menu and select "Earthcentric." Now the motions are not so simple! The Sun makes a regular loop around the Earth, but the other planets have more complicated motions. It is easy to see why it was thought from the time of the ancient Greeks until about 1600 AD that the planets move on epicycles – circles on circles – as they orbit the Earth.

As seen from Earth, each planet repeats a regular pattern in the sky. This cycle can be thought of as beginning when the Earth, Sun and the planets are all in a straight line, with the planet on the opposite side of the Sun from the Earth (this alignment is called *superior conjunction*), and ends the next time they come together in the same

# Kepler's Laws

Prior to the work of Johannes Kepler (1571 - 1630), no one understood how the planets orbit the Sun or the relationships between their orbits. Kepler had the good fortune to inherit the exceptionally precise planetary observations of his teacher, Tycho Brahe, and they allowed him to devise his three "laws of planetary motion." They summarize how the solar system is put together and how it operates. In simple terms, the laws state:

1. Planetary orbits are ellipses with the Sun at one focus.

2. Planets orbit fastest when closest to the Sun.

3. The length of a planetary year is related to its distance from the Sun by a simple formula.

Before Kepler, planet motions were a mystery; after Kepler, the planets' positions could be predicted accurately far into the future.

alignment. The length of this cycle depends not on the period of the planet's orbit, but on the relative motions of both the Earth and the planet about the Sun. Mars actually takes the longest to complete its cycle, about 26 months from superior conjunction to superior conjunction. During this cycle, a planet's phase and brightness will both vary, as will the best time to observe the planet. But how much these quantities vary depends on the particular planet. So let's take a look at the cycles of several different planets now.

## The Inferior Planets

We'll begin with the two *inferior planets*, to use a term that sounds disparaging but that is only a holdover from an era when all that distinguished one planet from another was its orbit. The inferior planets are Mercury and Venus, and they stay in the vicinity of the Sun as seen from the Earth. They can appear to the east (or left) of

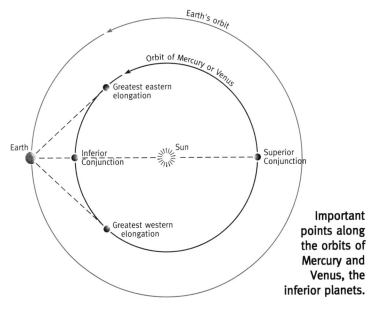

Important points along the orbits of Mercury and Venus, the inferior planets.

# Planetary Transits

When Venus or Mercury is at inferior conjunction, it may make a *transit* in front of the Sun. Planetary transits are actually a type of partial solar eclipse! Mercury and Venus are too small to eclipse the Sun when they move in front of it, and we see them cross the Sun's surface as slowly moving black dots. Mercury appears the size of a small sunspot and Venus the size of a large sunspot, so transits cannot be observed without properly filtered telescopes capable of fairly high magnification.

Just as solar eclipses happen only during "eclipse season" (see Chapter 4.1) when the Moon's orbit is so aligned that the Moon can pass in front of the Sun, so too planetary transits happen only when the

A planetary transit of Venus.

planet's orbit is aligned so that the planet can "eclipse" the Sun as seen from Earth. For this reason, transits are rare, and they occur in sets separated by wide intervals of time. Transits of Mercury occur alternately in May or November in 2003, 2006, 2016, 2019, 2032, 2039, etc. They last up to 7-1/2 hours. The only two transits of Venus in the 21$^{st}$ century are in June 2004 and June 2012. If you are lucky enough to watch a transit, you will see the planet slowly creep across the Sun from east to west, passing any sunspots that are present as it goes. Open the file "VenusTransit" and step through time to watch the transit of Venus in 2004. Change the location to see if it is visible from your home town.

the Sun, in which case they are visible in the evening sky; to the west (or right) of the Sun, in which case they are visible in the morning sky; behind the Sun, in which case they cannot be seen (superior conjunction); or at *inferior conjunction* when they lie between the Earth and Sun (which no other planets can be), and in which case they also cannot be seen. The ease with which these planets can be viewed is in direct correspondence with their angular separation from the Sun. The farther a planet appears from the Sun, the longer it is above the horizon during the night. Venus gets far enough from the Sun to be conspicuous much of the year, but Mercury never travels far from the Sun, so it usually can't be seen at all.

**Mercury** never strays more than 28° from the Sun. As seen from the Earth's equator it never sets more than two hours after the Sun or rises more than two hours before it. This number is smaller the farther you are from the equator, making Mercury even harder to see from mid-latitudes. The bottom line: Mercury is seen only near the horizon during twilight. It is best to look for it with binoculars and it is most easily found when the thin crescent Moon is near it to guide the way.

Open the file "Mercury" to follow Mercury through several orbital cycles, beginning with the planet at superior conjunction (on the far side of the Sun) on May 18, 2006. The orbital path which Mercury makes around the Sun is shown. Mercury is so nearly in line with the Sun that you will have to zoom in to see it. In reality, it rarely is **exactly** behind the Sun (just as it rarely passes exactly in front of the Sun) because its orbit is tilted 7° to the ecliptic. This is more than any planet but Pluto. Mercury is almost invariably north or south of the Sun when in conjunction with the Sun. Center and zoom in on Mercury to confirm that it is a tiny but full disk (you will have to zoom in very close, to less than 1'), and then zoom back out to the standard 100° field of view.

Step through time in one-day intervals and watch Mercury swing around to the left (or east) of the Sun. Its angular separation from the Sun increases daily until it reaches its maximum angular separation from the Sun on June 23. This is called its ***greatest eastern elongation***, which can be as little as 18° or as much as 27°. Zoom in on Mercury to verify that it is at third-quarter phase. The planet is so tiny – typically 5 arcseconds in diameter – and so near the horizon that its phase is hard to see even with a good telescope. Restore the field of view to 100° and change the time to 8 P.M. to see that it shines near the western horizon. Change the time step to 3 minutes and step forward in time to watch it set at 9:43 P.M. Mercury is visible only during a narrow window of time, after the sky becomes dark enough to see it, but before it sets.

Change the time back to 12:00 P.M. and the time step to 1 day. Start time running forward again. Mercury remains near its greatest eastern elongation for a few days, and its position in the sky changes little because it is moving nearly directly towards the Earth then, although it often moves a short distance north or south (in this instance) in a shallow arc. Then Mercury begins to move westward toward the Sun again. When it passes the Sun, it is on the near side of the Sun and much closer to the Earth than when on the Sun's far side, and Mercury's apparent speed across the sky is greater than when it was on the far side of the Sun. It is at inferior conjunction, when it is most nearly between the Earth and Sun, on July 18, and this is the date when it officially passes from the evening to the morning sky. It is now at its new phase (zoom in to verify, and then zoom out again), but you won't see it in the sky because it rises and sets with the Sun. A week later it reappears as a "morning star" near the eastern horizon during dawn. Telescopically it is a large (by Mercury standards) thin crescent. It climbs higher each day until it reaches its ***greatest western elongation*** on August 7. This coincides closely with its earliest rising time and its best pre-dawn visibility. At its greatest elongation it is at first-quarter phase. Mercury then moves more slowly back towards the Sun and disappears before it is in line with the Sun at its superior conjunction on September 1, when it is full again.

Mercury reveals little through a telescope. Ironically, it is one of the three planets whose surface is "visible" (Mars and Pluto are the others; the rest of the planets are shrouded by clouds), but it is too small and too distant for any surface features to be seen. Only its phase might be seen, and although it changes phase rapidly, the blurring effect of turbulence in our atmosphere (a problem with all objects observed at low altitude) makes it hard to tell what phase it is even through a good telescope.

**Venus** goes through a cycle similar to Mercury's. Open the file "Venus" and step through time in 3-day steps to follow a typical cycle of Venus, beginning at superior conjunction on October 27, 2006. Note how the orbital path of Venus takes it much farther from the Sun than Mercury's path does. Zoom in at any time to see the phase of Venus. When it is on the far side of the Sun it is small (about 10 arcseconds in diameter) and full. The Earth and Venus are moving around the Sun at similar speeds (30 vs. 35 km per second, or 18 vs. 22 miles per second respectively), so Venus increases its angular separation with the Sun only slowly. When it is 10° or more from the Sun, it can first be seen as the "evening star" low in the west after sunset. Its great brilliance makes it easy to spot despite its low height. It is now waxing gibbous. Months later it reaches its greatest eastern elongation (June 9, 2007), which can be as great as 47° from the Sun. At this time in its cycle, Venus sets after 11 P.M. at mid-northern latitudes and remains beautiful in the west long after twilight ends. It remains near its eastern elongation for weeks while it shines brightly late into the evening; it is then heading towards the Earth and moving very slowly against the stars. During the next several weeks it catches and then passes the Earth on an inside orbit, quickly moving toward inferior conjunction with the Sun (August 18, 2007). During these final weeks of its evening appearance, it becomes an increasingly thin and increasingly large crescent up to an arcminute in diameter – so large that its phase can be seen in very good tripod-mounted binoculars. It then seems to drop out of the evening sky, and in one month its visibility drops from very conspicuous to hard to see.

After passing inferior conjunction, Venus quickly reappears in the morning sky. It gains altitude rapidly day by day until it reaches its greatest western elongation (October 28, 2007), when it is half-full. Its disappearance from the morning sky is as slow as its appearance in the evening sky because it is on the far side of the Sun and its motion relative to the Sun is slow.

Venus is by far the brightest planet. Its brilliance comes from

## Galileo and the Phases of Venus

Galileo discovered the phases of Venus with one of his first telescopes, demonstrating that Venus orbits the Sun, rather than the Earth. If Venus orbits the Sun, so could the Earth, and this was important proof of the Copernican theory of a sun-centered solar system, which until then had been a mathematical curiosity. Galileo published his observations in code to establish the priority of his discovery until he could confirm it. When Galileo discovered the phases in October 1610, Venus was near full and quite small, and it was not until months later that Galileo could easily discern its non-round shape.

thick and highly reflective clouds that permanently shroud the planet. The surface features of Venus cannot be seen through any telescope. In fact, the clouds themselves are featureless in wavelengths visible to the eye. The planet is so brilliant it can be seen during the daytime if you know exactly where to look.

# The Superior Planets

The other planets, Mars through Pluto, remain outside the Earth's orbit and are called *superior planets*. They are only superior in an orbital sense. Unlike the two inferior planets, the superior planets never come between the Earth and Sun. Also unlike the inferior planets, whose visibility is limited to the hours just before sunrise or just after sunset, the superior planets can lie opposite the Sun and can be visible at any time of the night.

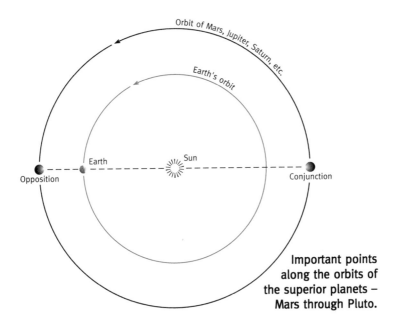

Important points along the orbits of the superior planets – Mars through Pluto.

Open the file "**Mars**" to follow one Martian cycle. We begin with the planet at superior conjunction on the far side of the Sun on July 1, 2000. Mars is almost 1° north of the Sun. Zoom in to see that it is at full phase – but the superior planets are always near full and can never be seen as crescent. Zoom back out and step forward through time in 1-day intervals. Mars is moving very slowly eastward against the distant stars, but the Sun is moving faster and leaves Mars behind. Unlike the inferior planets, which move faster than the Sun and reappear east of the Sun in the **evening** sky after their superior conjunction, Mars loses ground and reappears west of the Sun in the **morning** sky. The relative speed of the Sun and Mars is very low, and it takes at least a month for Mars to become far enough west of the Sun to be visible above the horizon in morning twilight. Because it is on the far side of the Sun, it is tiny as seen through a telescope and not especially bright at 2nd magnitude, so its cycle does not begin dramatically.

On January 1, 2001, change the time to 4 A.M. Resume running forward through time and notice that each morning Mars is higher at 4 A.M. than the morning before. This is because it rises a few minutes earlier each day. Notice that Mars is a few degrees north of the red star Antares, the "rival of Mars," early in March.

Around March 15, change the time to 2 A.M. and resume stepping forward through time to notice an interesting change in the motion of Mars. In April, 2001, Mars begins to slow noticeably in its eastward motion against the stars, and on May 15 it stops, is briefly stationary, then reverses its course and begins heading westward in *retrograde motion*. The planet does not actually move backward, of course; it is an optical illusion, but an effective one. During the several months that the Earth, which is traveling faster on an inside path, takes to pass Mars, Mars appears to move backwards against the stars. It is the same effect you get when you pass a slower moving car on the highway – in the time it takes you to pass it, the other car seems to move backwards in relation to the distant trees and hills, although both of you are moving forward at different rates.

Mars moves westward in retrograde motion at an accelerating rate through the summer. The Earth passes Mars on June 13, 2001, and then the red planet is at *opposition*; it lies opposite the Sun. Mars rises at sunset and is visible all night long. It is at its closest, 67.3 million kilometers (41.8 million miles) and at its brightest a week later. (The date of opposition and closest approach would coincide if the two orbits were perfectly round, but their non-circularity causes these dates to differ by a few days.). At this time, Mars is about four magnitudes brighter than it was at superior conjunction, or 40 times brighter! Mars and Venus vary in brightness much more than any other planets. You should be able to figure out why this is so.

Following opposition, Mars begins to fade as the Earth leaves it behind. It sets earlier each day. It continues its westward, retrograde motion until July 19, 2001, and then it seems to stop against the stars again and resume its normal eastward motion. Mars retrogrades toward Antares during early July, but reverses itself before reaching it. On July 1, 2001, change the time to 10 P.M., and resume running forward through time.

Zoom in during the late fall of 2001 and notice that Mars is nearly full, but not 100% so. This gibbous shape is as far from being full as an outer planet can be.

Mars moves eastward at an accelerating rate, crosses the ecliptic on February 10, 2002, and moves into superior conjunction with the Sun again on August 10, 2002. If you want to follow the path of Mars until it is again at superior conjunction, you will have to set the time back by a few hours in March or April 2002.

Each time the Earth passes Mars, Mars is not equally close because the two planets' orbits are not circular. The closest oppositions in the first third of the century come in August, 2003; July, 2018, and September 2035.

Telescopically, Mars is a challenging object. Even when closest to Earth it is surprisingly – and disappointingly – small, and high magnification on a good telescope with a steady atmosphere are required to see much. Practice is essential too – you will not see much at first glance, even on an ideal night. Allow time for your eye to pick out subtle details. The polar caps are bright and their whiteness contrasts well with the orange deserts that make up most of the planet's surface. The giant Hellas basin, when filled with white clouds, can rival the polar caps in brilliance. Indistinct dark markings are visible under ideal conditions and make the Mars aficionado wish for a larger telescope so as to see more.

Mars rotates in 24 Earth hours and 37.5 Earth minutes, so you can watch it rotate significantly during the course of an evening. Each night we see it turned slightly from its orientation at the same time on the previous evening. Open the file "MarsZoomed" and step forward at one day intervals to watch new features rotate into view while others rotate out. Each night surface features are offset by 37 minutes from the previous night, and during the course of a month you can see the entire surface. You will also notice Mars' two moons (Phobos and Deimos) come in and out of view. These moons revolve very rapidly – Phobos takes seven hours to complete an orbit of Mars, while Deimos takes 30 hours. These moons are visible only in very large telescopes.

**Jupiter's** path resembles Mars' path (and all the other superior planets), although its cycle is shorter. Open the file "Jupiter." We begin with it on the far side of the Sun on July 19, 2002. Step forward through time at one-day intervals, and change the time of day when necessary in order to keep Jupiter above the horizon. The Sun seems to race away, leaving it behind, and Jupiter first appears in the morning sky less than a month later, where its great brightness (second only to Venus) makes it relatively easy to see. It rises four minutes earlier each day (the same as the stars, since Jupiter's motion against the stars is so very slow) and eventually it rises before midnight and moves into the evening sky (October 26, 2002). On December 5, 2002, the Earth begins to catch Jupiter, and Jupiter begins its westward retrograde motion. The Earth and Jupiter are closest on February 2, 2003, which is also when it is at opposition (to within a few days). Jupiter ends its retrograde motion on April 5, 2003, resumes moving eastward, and is still moving eastward when it disappears into the glare of the Sun in August. It is at superior conjunction on August 21, and the 13-month cycle is completed.

Jupiter's great distance means that its size and brightness change little during its orbit, and it remains impressively large compared to the other planets. Even a small telescope will show a few cloud bands aligned parallel to its equator, and perhaps the Great Red Spot, which is a storm in Jupiter's atmosphere. A good telescope will show more features than a person can sketch. The planet's rapid 10-hour rotation allows you to see the entire surface over the course of one long night. The file "JupiterZoomed" shows the planet's rotation at 5-minute intervals over a long winter night.

Visible in this file are Jupiter's four major moons. These moons are a constant delight to amateur astronomers, and Starry Night will show the four major ones, which were discovered by Galileo. They can be seen in any telescope and even in binoculars. They orbit the planet so quickly (1.68 days for Io to 16.7 days for Callisto) that they change position night by night.

The gas giant Jupiter. The Great Red Spot is visible on the lower right.

Their motion can be seen almost minute by minute if they are next to the planet or to each other. Amateurs with decent telescopes enjoy watching the moons pass in front of Jupiter, cast their shadows on Jupiter (the shadows are dark and contrast well against the bright white clouds), disappear in eclipse in the planet's shadow, become occulted by the edge of the planet, or even (very rarely) eclipse each other. Jupiter's moons eclipse and occult each other when the plane of their orbit is seen edge-on to our line of sight, and this happens in June 2003, April 2009, Nov. 2014, March 2021, and Oct. 2026. Change the date in the "JupiterZoomed" file to the current date, if Jupiter is visible tonight, to see the changing configuration of its moons.

The motions of **Saturn** and the other outer planets are similar to Jupiter's and will not be treated here. You can use Starry Night to follow their motions many centuries into the future. Saturn, however, has rings and moons that set it apart.

Saturn's rings are one of the spectacles of the night sky when seen through a decent telescope. Several rings orbit the planet, nowhere touching it, and they have different brightnesses and widths. The rings are composed of grains of ice and rock. The planet itself is quite bland, but the rings make up for the lack of clouds.

Saturn's rings wrap around the planet's equator. Saturn's pole is tilted relative to its orbital plane, and as the planet orbits the Sun we see its rings at a constantly-changing orientation that follows Saturn's 30-year orbital period. When Saturn's pole is tilted the greatest amount towards or away from the Earth, as it is in 2002 and

**The rings of Saturn.**

2017, we see the rings "open" to our view with a tilt of 27 degrees; they are wide and bright. Halfway between these dates, the rings are briefly edge-on to our view, and then they almost disappear. They are edge-on in September 2009 and March 2025, but the planet is behind the Sun on both these occasions and we will miss the novelty of seeing Saturn without its rings. We will not actually see a ringless Saturn until 2038. Open the file "SaturnRingTilt" and run forward through time to see how the appearance of Saturn's rings changes over time.

Saturn's moons are more challenging than Jupiter's major four, but huge Titan can be seen in any telescope and even a modest telescope will show one or two others. Ninth-magnitude Titan orbits Saturn in 16 days, and it lies five ring-diameters from Saturn when east or west of the planet. Rhea, Dione, and Tethys are between 10th and 11th magnitude and orbit even more quickly, closer to the planet.

**Uranus** and **Neptune** are bright enough to see in binoculars or small telescopes, and Uranus has been spotted with the naked eye, but the trick is to know where to look. Their tiny greenish disks distinguish them from stars under high magnification (which is how William Herschel discovered Uranus), but they show no surface features and their moons are visible only with large telescopes. They move slowly from constellation to constellation and hold little interest for amateur astronomers. Use Starry Night to see if they are visible tonight.

**Pluto** looks like a 14th magnitude faint star, and it can be spotted only in a very large amateur telescope. Pluto is much fainter than the faintest stars Starry Night can show, so if you wish to find Pluto with your giant telescope, use Starry Night to find Pluto's coordinates and then mark its position on a detailed star chart. Confirm that you have found it when you record its slow movement against the stars from night to night.

# Conjunctions

As the planets orbit the Sun and move across the sky, one occasionally passes another, and when this happens it is called a *conjunction*. Conjunctions happen almost every month and are not rare, but some are more interesting than others. Some are spectacular.

A conjunction is defined as the moment when two planets have the same ecliptic longitude (see the section on "Coordinate Systems" in Chapter 2.2 for a definition of ecliptic longitude). Alternately, it is the moment when two objects have the same right ascension, which is slightly different. Neither is necessarily the exact moment when they are at their closest. A conjunction looks different from different locations on Earth because of parallax. The conjunctions (superior and inferior) that we looked at in the above section all involved the Sun and another planet, but the most interesting conjunctions do not involve the Sun, but instead two or more planets, or one or more planets and the Moon.

Mercury and Venus move so quickly that they are involved in most conjunctions that occur. Slow-moving Saturn, in contrast, seldom passes another planet, but is itself often passed. The Moon is in conjunction with each planet once a month, and if it comes near the bright planet Venus, Mars, or Jupiter, it can be a spectacular naked-eye sight. You can discover conjunctions to your heart's content by running Starry Night through time.

A conjunction is interesting to amateur astronomers when the two objects come close enough to each other to be visible at the same time in a pair of binoculars or, exceptionally rarely, through a telescope. Even rarer is an occultation of one planet by another. None will occur within the lifetime of most people now alive. At 12:45 on November 22, 2065, the centers of Venus and Jupiter will be 16 arcseconds apart and the northern edge of Venus will pass in front of Jupiter! Mark it on your calendar.

Three or more planets cannot be in conjunction simultaneously, although the word is sometimes misapplied when several are close to each other. Such a grouping of planets is called a *massing*, and you can discover massings with Starry Night. The best in history was in early 1953 BC, and it was profoundly important to the ancient Chinese, who recorded and remembered it. There are files with this chapter for planetary massings on September 29, 2004 and December 10, 2006 ("Massing 2004" and "Massing 2006" respectively).

A popular misconception is that all the planets sometimes line up like billiard balls in a row, and dramatic images of such an alignment are even seen on book covers. The reality is different.

Jean Meeus, the foremost thinker about planet alignments, has calculated how close the planets come to

**The massing of all five naked eye planets in 1953 BC.**

being in a line. He finds that, between the years 3100 BC and 2735 AD – a time span which includes all of recorded history – the minimum separation of the five naked-eye planets was 4.3° on February 27, 1953 BC. The best groupings closest to the present are: April 30, 1821 (19.7°); February 5, 1962 (15.8°); May 17, 2000 (19.5°); and September 8, 2040 (9.3°). The 2040 grouping, which includes the crescent Moon at no extra charge, will be spectacular.

If you consider the three outermost planets too, all the planets never line up.

A *triple conjunction* happens when a planet makes its retrograde loop near a star; the planet passes that star once in its normal forward motion, a second time when in retrograde motion, and a third time when it resumes its forward motion. Each outer planet is in triple conjunction with all the stars within the bounds of its retrograde loop, but occasionally a bright star (or even another planet!) is within that loop. (The object does not have to be within the loop itself, but within the limits of the loop in Right Ascension or ecliptic longitude.) Open the file "VenusConjunction" to view a triple conjunction of Venus and the star 41 Piscium. Step through time and watch how Venus passes this star 3 times: in February 2001, in April 2001, and finally in May 2001. It is believed by many that a triple conjunction of Jupiter and the star Regulus in the years 3 and 2 BC is the historical basis for the Star of Bethlehem.

Some people whose knowledge of astronomy is limited are concerned that when planets align, their gravity is somehow magnified and they exert a major pull on the Earth. The gravitational or tidal force of a planet is completely insignificant, and even when

## Star of Bethlehem

The famous Star of Bethlehem, seen so often on Christmas cards and topping off Christmas trees, may have been a conjunction of planets. Astrologers of the time looked to the sky for omens, and one of the best was a conjunction of bright planets. In the summers of 3 and 2 BC – the years when Jesus is most likely to have been born – there were two very close conjunctions of Jupiter and Venus (Venus actually occulted Jupiter as seen from South America in the second conjunction) and three conjunctions of Jupiter and the star Regulus. It is plausible that the wise men, who were astrologers, would have interpreted so interesting a series of conjunctions as the fulfillment of ancient prophecies.

several are massed together it has no effect on our planet or on ourselves. Planetary alignments do not cause cosmic disasters.

## Minor Bodies: Asteroids and Comets

On New Year's Day, 1801 – the first day of the 19th century – the Italian astronomer Giuseppe Piazzi discovered a new "planet" between the orbits of Mars and Jupiter. It was much smaller and fainter than the other known planets, so it was called a ***minor planet*** or ***asteroid*** (from its starlike appearance). Since then, over 10,000 asteroids have been discovered and they are a fascinating group of objects.

**Asteroids** are fragments of small planets that were shattered in collisions with each other early in the history of the solar system. Jupiter's gravity prevented the formation of one big planet inside its orbit, and the many asteroids are the result. Most asteroids orbit within the asteroid belt – a wide zone between the orbits of Mars and Jupiter – but many others have been knocked out of the asteroid belt during subsequent collisions and wander throughout the solar system.

The largest asteroid is Ceres, which has a diameter of 930 kilometers (580 miles). It and all the other **large** asteroids remain safely within the asteroid belt, but small ones, which can drift throughout the solar system, can strike the Earth. When one does, it falls through the atmosphere as a ***meteor***, and if it survives to reach the ground it is called a ***meteorite***. Meteorites are fragments of shattered planets. The impact of a giant asteroid fragment (or comet fragment) is widely credited for the extinction of the dinosaurs 65 million years ago. Most meteorites are stone, but a few are made of iron.

The largest asteroids can be seen with binoculars or a small telescope **if** you know where to look. Starry Night will show asteroids

# Meteors and Meteor Showers

Most meteors come from comets. Comets shed dust as their ices evaporate, and this dust never returns to the comet. The dust drifts around the Sun, following the orbit of the comet. If a dust particle is swept up by the Earth, it falls through the atmosphere towards the ground. Friction with air molecules heats it to incandescence and it bursts into flames; we see it as a meteor. Alternate popular terms are "falling star" or "shooting star."

If the Earth passes through or near the orbit of a comet, we pass through a region filled with dust, and meteors fall by the hundreds. A shower happens at the same time each year, when the Earth returns to that part of its orbit. The shower can last for a few hours to a few weeks. Listed below are the best meteor showers, their peak dates, and the number of meteors per hour an observer in a dark location might optimistically see:

| Shower | Date | Hourly Rate |
| --- | --- | --- |
| Quadrantid | January 3 | 85 |
| Eta Aquarid | May 5 | 30 |
| Delta Aquarid | July 29 | 20 |
| Perseid | August 12 | 100 |
| Orionid | October 22 | 20 |
| Geminid | December 14 | 100 |
| Ursid | December 22 | 45 |

The Eta Aquarid and Orionid showers come from Halley's Comet on the inbound and outbound legs of its orbit respectively. The Geminids come from the asteroid Phaethon.

in the Planet List, and you can center and lock on them like any other object. Open the file "Asteroids" and start time running forward to see the orbits of 5 major asteroids, all of which are between Mars and Jupiter in the asteroid belt. Unlike the orbits of the planets, the orbits of these asteroids (particularly Pallas) are tilted at large angles to the plane of the ecliptic. The thrill in observing asteroids is simply to find and track them. Dozens are within the range of amateur equipment.

The planets and minor planets circle the Sun with great regularity and we can know exactly where they will be thousands of years in the past or future. This is not true for comets. Most of them have not yet been discovered – and won't be for thousands of years!

**Comets** were born in the outer fringes of the solar system, and that is where nearly all of them still reside. There is a huge reservoir of billions of giant ice balls beyond the orbit of Neptune, of which Pluto is the largest member. If these ice balls stay there, we will not know of them (except by diligent searches conducted with large telescopes) and they will not bother us. However, occasionally one is deflected out of its distant but stable orbit into a new orbit that brings it into the inner part of the solar system, and then we see it as a comet. Comet Hale-Bopp, for example, was first sighted near the orbit of Jupiter in 1995, about two years before it was at its brightest and closest to the Sun.

A comet is a chunk of ice and dust only a few kilometers across. The ice is mostly frozen water and carbon dioxide and the dust is simple silicates. When far from the Sun, the ice is frozen and the comet is too small and too faint to be seen. But as the comet approaches the Sun, it warms. Sunlight thaws the ice, which evaporates (it does not melt – there are no puddles in space!), carrying with it the dust that was embedded in it. The gas and dust form tails that can be long and beautiful. Sunlight causes the gas to fluoresce while the dust merely reflects sunlight, and the gas tail often has a different shape and color than the dust tail. A comet is brightest when it is closest to the Sun, which is when it is warmest, and that is when it is traveling the fastest. If it is near the Earth at the same time, it can zip across our sky at the rate of several degrees a day. As it leaves the Sun's vicinity, it cools, becomes fainter, and slows down. It may not return for hundreds or many thousands of years. Some comets have been captured into short orbits and appear every few years, but these comets have had their ice evaporated after spending so many years near the Sun and they are exhausted, and all are faint, with the exception of one – Halley's. Halley's Comet is the **only** bright comet that returns often enough for each person to have a chance to see it, and many people have. It returns approximately every 76 years, most recently in 1986. Like any other comet, Halley's loses ice through evaporation every time it approaches the Sun, and it is not as bright now as it was the first time it passed through the inner solar system.

Several prominent comets are shown in the Planets List in Starry Night. If you choose "File | Preferences" from the Starry Night menu and press the "Update Now" button, Starry Night will automatically update your comet data to add any prominent new comets in our Solar System. As with any other object in the Planets List, you can then center and lock on a comet to find its place in the sky.

Starry Night's representation of Comet Hale-Bopp. The gas tail, is quite narrow, while the dust tail is more widely scattered.

# Section 5

# Deep Space

ONCE we get out of the solar system, the motion of objects in the heavens becomes a lot less confusing. Stars, galaxies and other deep space objects are so far away from us that their position along the celestial sphere will change little during our lifetime. This sense of permanence allows us to associate patterns of stars as constellations, and describe where a deep space object is with regards to these constellations. Chapter 5.1 describes the various deep space objects, Chapter 5.2 gives more background on the constellations, and Chapter 5.3 concludes with a tour of the highlights of the constellations, covering many fascinating deep space objects along the way.

# STARS AND GALAXIES

## Stars

We learned in Chapter 1.1 that we classify a star's brightness by a number called its magnitude. A star's magnitude tells you how bright that star appears to your eye as seen from Earth, but it tells you nothing about its intrinsic brightness (the amount of light it emits). For that reason, magnitude is more properly called *apparent magnitude*. To know a star's intrinsic brightness, you must know its distance from Earth.

Obviously, the closer a star is, the brighter it appears, and many of the brightest stars in our sky are among the closest. Sirius, Procyon, Altair, and Alpha Centauri are excellent examples of very close and very bright stars. But if you could line all the stars up side by side, at the same distance from Earth, you would see that some outshine others by many thousands of times. The *absolute magnitude* compares how bright different stars would appear if they were all the same distance from Earth. As with apparent magnitude, a lower number means a brighter star, and a difference in absolute magnitude of 1 corresponds to a brightness difference of 2-1/2 times. The Sun has an absolute magnitude of 4.8, which is about average for stars. Faint dwarf stars have absolute magnitudes as high as 15,

## Determining the Distance to the Stars

We find the distance to the nearest stars by measuring their *parallax*. Parallax is the shift of a nearby object against more distant objects when seen from two positions. A simple demonstration of parallax is to view an outstretched finger on your hand, first with your left eye and then with your right eye; notice how your finger "shifts position" relative to the background as your viewpoint shifts from eye to eye. If you move your finger farther away and perform the same experiment, you will notice that the position shift is smaller. The same idea applies to measuring astronomical parallax.

Astronomers record a star's position, and then record

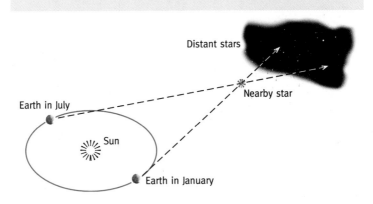

Distant stars

Nearby star

Earth in July

Sun

Earth in January

while bright giant stars like Antares and Deneb have absolute magnitudes as low as -5.

The stars we can see from Earth are not representative of all stars. We see the true powerhouses when we gaze into the night sky, as almost all the stars visible to the naked eye have a lower absolute magnitude (are intrinsically brighter) than our Sun. The

it again six months later when the Earth has moved halfway around the Sun and is 300,000,000 km from where it was at the time of the first observation This will cause the star's position in the sky to shift, and the amount of this shift is the parallax. More distant stars will shift less, therefore having a smaller parallax. The technique works for relatively nearby stars, but distances to distant stars, star clusters and galaxies are determined through indirect means because these objects are so far away that their parallaxes are too small to measure accurately. This means that distances to these objects must be used with caution. You will see wildly different distances quoted for a star cluster, for example, each of which represents the best effort of an astronomical research project. The latest and presumably most correct values are used in this book, but they too are subject to revision and should not be taken literally

Parallax is usually expressed in arcseconds, and this explains the origin of a strange-sounding word. A *parsec* is the distance an object would have to be if it had a parallax of one arcsecond, and one parsec equals 3.26163 light years. The parsec is the most commonly used unit of distance for astronomers.

most luminous stars are so energetic that they shine like lighthouses across the gulf of space and can be seen from enormous distances, while a typical star shines brightly only in its own neighborhood and the faintest are hard to see even from nearby. The common types of stars exist in abundance (in fact, our Sun is intrisically brighter than most stars), but for the most part are invisible without a telescope.

Look closely, and notice that some stars have color. Most stars appear white, but a few have a slight blue or orange hue. Star colors are not saturated, and your eye is not very sensitive to colors in low-light level objects, so star colors are subtle. A reflecting telescope will enhance star colors, although inexpensive refracting telescopes often cause stars to exhibit false colors through incorrectly refracted starlight. Deneb, Rigel, and Spica are bluish, while Antares, Aldebaran, and Betelgeuse are slightly orange in color. The color of a star tells you its temperature, and its temperature tells you what kind of star it is. Red stars are cool and blue stars are hot. Bright red stars (which actually look orange or even yellow) are "red giants" or "supergiants." These giants are hugely swollen stars with bloated atmospheres. They can be so large that, were one placed at the center of our solar system, Mars would orbit beneath its surface! Blue stars shine brightly not because of their size, but because of their intense heat. Their high surface temperature of 25,000° C for a blue giant vs. 3,500° C for a red giant (45,000° F vs. 6,300° F) gives them luminosities of 10,000 times the Sun or greater. Other red stars, called "red dwarfs," are the most common type of star, but all are so intrinsically faint that none, including the closest, can be seen without a telescope. By playing with the sliders in the "Brightness and Contrast" option in the "Sky" menu, you can have Starry Night draw the stars in exaggerated shades of blue and red.

Slightly more than half of the stars you see sit alone in space (disregarding any planets they may have). The rest are accompanied by a companion star or stars, and these are called *double stars*, "triple stars," or even "multiple stars." They are very popular targets for amateur astronomers with telescopes of all sizes, as the telescope often reveals two or more stars where the unaided eye sees only one.

Most double stars are true *binary* systems, and the two stars are in orbit around a common center of gravity. The star info window in Starry Night will tell you if a star is a double, and will give the angular separation between the two components. Open the file "Mizar" to see a famous multiple star. You begin by looking at the star Mizar (in the handle of the Big Dipper) with a regular 100° field of view. Use the Zoom button to zoom in. Once your field of view is down to about 20°, you will see that Mizar is actually two stars, Mizar and Alcor. If you let the cursor hover over each star, you will note that their distances are similar, indicating that they are most likely gravitationally bound. Keep zooming in. When you reach a field of view of about 20', you will see Mizar again split into two stars, officially called Mizar A and B. These stars also have similar distances and are part of the same star system. Recent observations have revealed that all three of these stars have faint companions, so there are actually six stars in this system!

Occasionally two stars happen to line up as seen from Earth, even though one lies far beyond the other, and such pairs are called "optical doubles." Open the file "AlphaCap" and zoom in to see this optical double in Capricorn split into two distinct stars. The brighter one, Algedi is 109 light years away, while the fainter star is almost 700 light years distant! Clearly this is just a chance alignment, not a true multiple star system.

Some amateur astronomers enjoy finding and monitoring *variable stars* – stars that change their brightness. Some variable stars, like Delta Cephei, are regular and predictable, pulsating like clockwork in a cycle that takes a few days. Others, like Mira, follow a rough pattern that is similar year-to-year but that can hold surprises.

Some variable stars have huge changes in their brightness from maximum to minimum. Mira, for example, goes from a magnitude of about 3.5 at maximum to 9.5 at minimum, which means a change in brightness of a factor of 200! The star info window in Starry Night will tell you if a star is variable. One of the areas where amateurs can make a serious research contribution is by monitoring variable stars, systematically recording their brightnesses, and forwarding the observations to a centralized authority like the American Association of Variable Star Observers (AAVSO). There are so many variable stars that it is impossible for professionals to make all these observations, so they rely heavily on the AAVSO data.

## Naming A Star

You cannot name a star for a friend or loved one. Although several companies will offer to name a star and to "register" it in an "official document," for a fee, such vanity registries have no legality. They are simply listings in the company's own book and no one else will recognize the star's name.

The responsibility for naming stars, as well as comets, newly discovered asteroids and moons, surface features on other planets, rests with a committee of the International Astronomical Union. Although they have been busy naming newly revealed craters on Venus, for example, they do not sell star names.

Star names sold by private companies have no official status, and these companies actually sell nothing more than pretty pieces of paper at exorbitant prices. Save yourself the expense and print your own equally valid certificates on your home computer.

# Star Names

The familiar star names that sound so foreign to our ears are, for the most part, Arabic translations of Latin descriptions. Arabic astronomers translated into their language the Greek descriptions of Ptolemy, and centuries later scribes in the Middle Ages copied and recopied the text, introducing errors while copying words they did not know, until the origins and meanings of some words became difficult to decipher. A few stars names are relatively modern and some were invented as recently as last century.

This representative list of common and unusual star names gives a feeling for how stars got their names.

Aldebaran: from Arabic for "the follower," as it follows the Pleiades Star Cluster across the sky.

Algol: from Arabic "al-gul" – the ghoul; the Arabs apparently noticed – and were bothered by – the way it varies its brightness.

Antares: from Greek "anti-Aries," meaning against Mars (or more colloquially "rival of Mars"). "Aries" is Greek; "Mars" is Latin.

Betelgeuse: a corruption (copying error) of "bad" from "yad al-jauza" – the hand of al-jauza, the Arab's "Central One."

Regulus: diminutive of king, (as in "regal"); named by Copernicus

Sirius: from Greek "serios" for searing or scorching

Thuban: corrupted Arabic for "serpent's head"

Vega: "falling" in Arabic, as the Arabs thought of it as a bird falling from the sky

Fainter naked-eye stars were given numbers or Greek letters according to schemes devised centuries ago by John Flamsteed and Johann Bayer respectively when the first detailed star charts were printed in book form. Starry Night displays common names as well as Bayer letters and/or Flamsteed numbers in the star Info Window. If a star is too dim to have a common name, a Bayer letter, or a Flamsteed number, it is identified in Starry Night by its Hipparcos (HIP) catalog number or its Tycho (TYC) catalog number. These two catalogs were the product of the Hipparcos project, a mission by the European Space Agency to calculate the distances to nearby stars by measuring their parallax (the star distances in Starry Night come from this project). The Hipparcos catalog has about 100,000 stars. Many other star catalogs exist, and other texts may refer to a star by its number from one of these catalogs.

## Star Clusters and Nebulae

Anyone who has attended a public "star party," where amateur astronomers conduct guided tours of the wonders of the sky with their telescopes, has seen at least a few star clusters and nebulae. They are indeed the showpieces of the night sky. Searching them out and examining them is one of the sublime and endless activities that delights amateur astronomers (and causes them to crave ever-larger telescopes).

Star clusters come in two varieties: ***open clusters*** (sometimes also called galactic clusters) and ***globular clusters***. The two have distinct appearances and histories. Clusters were born together out of enormous clouds of gas, and the stars of a cluster are all the same age.

The brightest open star clusters can be seen with the unaided eye. These include the Pleiades and Hyades clusters in Taurus and the constellation Coma Berenices. Several others – the Beehive in Cancer and the Perseus Double Cluster in Perseus – are visible to the unaided eye but are best in binoculars or low-power spotting scopes. (See the descriptions of these objects in the sections on their respective constellations in Chapter 5.3.) Hundreds more await amateurs with their telescopes. You can search for the most popular ones by name and display them in Starry Night.

Open star clusters take their name from their "open" or undefined shape. They lie in and near the arms of our Milky Way Galaxy. Some are forming still, many are quite young (astronomically speaking), and the oldest were born as much as a billion years ago. The stars in an open cluster are weakly held to each other by gravity, and the cluster eventually "evaporates" by losing its outermost stars, which escape into interstellar space. More than 1,000 are known, and they contain from a dozen to a few thousand stars. Each has its own characteristics, some have unusual appearances (like the Pleiades, which resembles a very little dipper), and no two look entirely alike.

The Pleiades, an open star cluster in the constellation Taurus.

Globular star clusters, in contrast, were born during the early days of the Milky Way and are as old as our galaxy itself. They are huge compact spherical balls of ten thousand to a million old stars, and at first glance they all look identical. A closer examination shows that they have varying degrees of compactness, but differences between them are indeed subtle. About 100 are known in our galaxy. They form a halo distributed around the center of the Milky Way, and they avoid the spiral arms. Because they pass near the center of the Milky Way at some point in their orbit, they appear concentrated towards that direction, which is in Sagittarius in the summer sky. None are very close to our solar system. Whereas many open star clusters lie a few hundred to a thousand light years from Earth, a typical globular cluster is tens of thousands of light years distant, and its stars are proportionately fainter. Only a small handful (such as M13 in Hercules and M22 in Sagittarius in the Northern Hemisphere and Omega Centauri in the Southern) are visible to the unaided eye, and they look like faint stars. Only through a telescope can their structure be seen. Open the file "M13" and then zoom in to see the structure of the Hercules Cluster.

A *nebula* is an enormous cloud of gas ("nebula" is Latin for cloud). Our Milky Way is permeated with gas, most of it hydrogen and helium, which is concentrated in its spiral arms. For reasons not completely understood, this gas can become concentrated into comparatively dense nebulae. If they are illuminated by nearby stars, nebulae shine brightly and are beautiful tenuous wispy clouds when seen through a telescope; if unilluminated, they appear as blotches of darkness silhouetted against the myriad stars of the Milky Way. The brightest, the Orion Nebula, is barely visible to the unaided eye as a "fuzzy star" in Orion's sword. It is impressive through any telescope if the sky is dark (being low-contrast objects, nebulae require a dark sky to be seen well). Several others that are pretty when observed with amateur equipment lie in Sagittarius. The most famous dark nebula is the "Coal Sack" in the Southern Cross. Nebulae are the birth places of stars, and several are giving birth to brand-new clusters of stars right now. Hot newborn stars then disperse the remaining nebulosity. When the Milky Way is much older than now, its gas will be exhausted and no new nebulae – and no new stars – will form.

M16, the Eagle Nebula

Another type of nebula, called a *planetary nebula*, is entirely different. A planetary nebula is a shell of gas expelled by an aged or dying star. The Dumbbell Nebula in Vulpecula and the Ring Nebula in Lyra are pretty sights in amateur telescopes and are popular showpieces at summer star parties, but most are faint and a challenge to find. Their confusing name comes from their superficial resemblance to the planets Uranus and Neptune as seen through a small telescope. Open the file "Orion Nebula" to see a true nebula, and then open "Dumbbell" to see a planetary nebula.

## Messier Objects

You have probably noticed that many of the objects we have been looking at have the letter 'M' followed by a number. These are objects in the *Messier* catalog. There are 110 objects in this catalog, which was compiled by astronomer Charles Messier in the 18th century. These objects are either star clusters, nebulae, or galaxies. Messier was a comet hunter, and his goal in making this catalog was to identify "fuzzy" objects which were commonly mistaken for comets! Today the catalog is more frequently used as a guide to some of the most beautiful objects in the sky. Because Messier was based in France, the catalog has a definite Northern Hemisphere bias, and many treasures in the Southern sky are missing from his catalog. Starry Night includes images of all the Messier objects.

# The Milky Way

Everything we have looked at so far in this book is inside our own galaxy, the *Milky Way*. Much as the two-dimensional plane of the ecliptic appears as a one-dimensional line when viewed from Earth, the three-dimensional Milky Way appears as a two-dimensional band of countless stars that wraps around the sky. This band appears brighter than the surrounding area, and its brightness comes from the combined light of billions of dim stars. The Milky Way is the largest and grandest structure in the sky.

A view of the Milky Way from Starry Night.

Open the file "Milky Way" to see the outline of the Milky Way across the southern part of the celestial sphere. (If you want to change the color of this outline to make it more prominent, choose "Sky | Sky Settings," then click on the color bar beside the words "Milky Way Color" to select another color.) Now choose "Sky | Milky Way" to turn off this outline. You can still see that the area of the sky inside the Milky Way has many more stars than the surrounding regions. Turn the Milky Way outline back on and scroll around to see what other constellations the Milky Way passes through. You can see that the Milky Way is not uniformly wide and bright. It is much brighter and wider in the portion that extends from Sagittarius south to the Southern Cross, and it is comparatively thin in the opposite direction, towards Perseus and Auriga. This asymmetry shows us the direction to its center, which lies about 28,000 light years from us in the direction of Sagittarius.

The Milky Way has a third dimension: depth. Because we are inside the Milky Way, we cannot view its entire extent. However, by mapping the distribution of stars and gas, scientists know that it would look like a giant flat pinwheel with a bright center and spiral arms made of stars and clouds of gas, all of which is surrounded by a halo of dispersed stars and unknown material. This structure is known as a ***spiral galaxy***. Open the file "M100" to see another spiral galaxy, one that looks very much like our Milky Way would from the outside.

The Milky Way is over 100,000 light years in diameter but only a thousand light years thick. It contains between a hundred billion and a trillion stars. The very center of the Milky Way seems to be inhabited by a giant black hole with the equivalent mass of 2.6 million suns. The black hole is surrounded by spiraling clouds of hot hydrogen gas. None of this can be seen by our eyes because of numerous dust clouds that lie between us and the center, but it has been detected and mapped by radio astronomers.

Surrounding the Milky Way's core is the central bulge, which is several thousand light years in diameter and which contains billions of stars. You can see part of the central bulge on a summer evening when you look at the Great Sagittarius Star Cloud.

Radiating from the center are a suite of spiral arms. No one knows what causes the arms to form or what maintains their structure. They are bright because they contain gas and the hot young stars that were recently born out of the gas. The Small Sagittarius Star Cloud is a small portion of the innermost arm, which is about 6,000 light years distant.

The spiral arms rotate around the center. At our distance from the galaxy's center, the Sun takes 230 million years to make one revolution – a period of time called a "galactic year." We are moving toward Cygnus as we orbit the Milky Way.

Our Sun lies on the inside edge of the Orion Arm. When we face Orion, we are looking at the nearest arm to us, which is part of the reason there are so many bright stars in Orion's vicinity. The Orion Nebula is near the arm's center. Beyond lies the Perseus Arm at a distance of about 7,000 light years; the Perseus Double Cluster is within it. Beyond lies the sparse outer edge of the Milky Way.

All the stars that make up the constellations inhabit a tiny portion of the Milky Way. We see just our neighborhood. The naked-eye stars are less than 2,000 light years from Earth (most are much closer), and this is less than 2% of the diameter of the Milky Way.

Binoculars (or a low-power wide-angle spotting scope) are the best tool for exploring the Milky Way. Take the time to explore it at leisure when you are away from city lights.

As much as we've learned about the Milky Way, serious questions remain unanswered. We do not know how it formed, how it acquired its globular clusters, the exact nature or history of the giant black hole at its heart, its precise shape (it seems to have a bar structure running through its center – the "Milky Way Bar"), how many spiral arms it has, or the nature of the "missing mass" that is important in holding it all together. Much remains to be discovered before we can claim to know it well.

# Other Galaxies

The Milky Way is our galaxy, but billions of other galaxies exist. You can see three of them with your eyes alone and hundreds with a decent amateur telescope.

Early last century galaxies were called "island universes," which evokes their size and isolation. Galaxies come in a variety of shapes and sizes, but they have in common that they are vast systems of many billions of stars, each with its own history. Galaxies are the fundamental building blocks of the universe, and they stretch away as far as modern telescopes can see – to distances exceeding 10 billion light years. Billions of galaxies are known.

The closest large galaxy to the Milky Way is the "great galaxy in Andromeda" or more concisely the Andromeda Galaxy. It is also the brightest galaxy visible from the Northern Hemisphere. It can barely be seen with the unaided eye on a dark autumn night, but it shows up well even in a small telescope. It lies about 3 million light years from Earth and is comparable to the Milky Way in size. (See the section on Andromeda in Chapter 5.3.)

Much closer but visible only from the Southern Hemisphere are the two Magellanic Clouds. These comparatively small galaxies are satellites of the Milky Way, and they orbit around it like galactic moons. They will probably collide with the Milky Way in the distant future and become absorbed into it, as has certainly happened to other small galaxies that approached too close to the Milky Way in the distant past. To the eye, they look like detached portions of the Milky Way. They lie a bit less than 200,000 light years from Earth.

Beyond are countless galaxies visible only with a telescope. They look like faint smudges of light, but each has its own character and its own orientation. The largest have either a spiral shape like our Milky Way, some with a bar running through their center, or they are elliptical superhuge balls of stars with little if any gas. A few are irregular in shape. The most common are small elliptical galaxies that resemble globular star clusters. It can be hard to distinguish a small *elliptical galaxy* from a large globular cluster. Open the file "Elliptical" to see M49, a good example of an elliptical galaxy.

The greatest concentration of nearby galaxies lies in the direction of Virgo. Our Milky Way is actually on the outskirts of

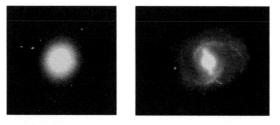

Two Messier objects which are close together in the Virgo Cluster. M87 (left) is an elliptical galaxy, while M91 is a spiral.

this huge cluster of galaxies whose center is 60 million light years from home. Two dozen Virgo galaxies are bright enough to see in large binoculars.

Many amateurs enjoy hunting down galaxies, trying to tease out their structures, and comparing them to each other.

# ABOUT THE CONSTELLATIONS

## Introduction

In this chapter we return to learn more about the constellations, the fundamental divisions of the sky.

Few constellations look like the animal or person they are named after, and you should not be frustrated if you cannot see a princess, bear, or winged horse in the sky. Only in a few cases can the stars be connected to be made to look like a man, a lion, or a swan. Many constellations were named in honor of heroes, beasts, and objects of interest to the people who named them, and not because of any physical resemblance. We do similar things today: the state of Washington looks nothing like the gentleman on the U. S. dollar bill, nor does the bridge in New York City. The exceptions are few, among them Orion, Scorpius, Gemini, and Taurus.

During the Renaissance, when star maps were designed to be beautiful as well as useful, they were filled with elaborate and colorful drawings that often ignored the background stars. Starry Night will show constellation figures in the classical style if you choose, but notice how poorly they fit the stars. Choose "Constellations | AutoIdentify" and scroll around to see these classical drawings for each constellation.

There is more than one way of drawing the connect-the-dot stick figures which we make to identify the constellations. H.A. Rey created a new set of figures in his book *The Stars: A New Way to See Them*, which remains in print five decades after it was first published. Rey designed his figures so that the pictures they made corresponded with the constellation names. These figures appeal to people who feel the need to find a one-to-one correspondence between stars and figures regardless of whether or not a meaningful fit can be made. In any case, there are no "official" or correct ways to connect the stars, and you are free to invent your own designs if you like.

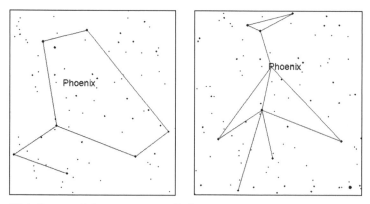

**Stick figures of the same constellation can vary dramatically. These drawings show the constellation Phoenix in both the standard figure (left) and that invented by H.A. Rey.**

## Asterisms

Some common star patterns are not actual constellations. These *asterisms* (from aster, the Greek word for star) can be part of one constellation or parts of two or more constellations. These are a few of the major asterisms; select "Constellations | Asterisms" in Starry Night to display them all.

**Big Dipper:** The seven brightest stars of Ursa Major.

**Great Square of Pegasus:** The three brightest stars of Pegasus and the westernmost bright star of Andromeda.

**Keystone:** Four medium-bright stars in Hercules that form a square.

**Little Dipper:** The seven brightest stars in Ursa Minor.

**Sickle:** Stars in Leo in the shape of a harvesting sickle (or a backwards question mark).

**Summer Triangle:** The stars Vega in Lyra, Deneb in Cygnus, and Altair in Aquila.

Constellations have both a formal Latin name (Aquarius, for example) and a common English equivalent (Water Carrier). This differs from star names, which are mostly Arabic, as you learned in Chapter 5.1. Some constellations pre-date the Latin-speaking Romans, but Latin was the language of scholarship until recently and the Latin names were standardized long ago. New constellations invented in modern times (such as Leo Minor) were given Latin names to conform with ancient custom. Choose "Constellations | Constellation Settings | Labels" to choose which style of name (or both) you want displayed in Starry Night.

See Appendix A for a list of constellations and their abbreviations.

At any given moment you can see half the sky – and approximately half the constellations. As the hours pass and the sky rotates overhead, constellations in the west set and are replaced by others that rise in the east. During the course of a long winter night you can see perhaps two-thirds of the sky, and during the course of a year even more. But – unless you live on the equator – some constellations remain permanently hidden from view. The far southern constellations are a mystery to us in the Northern Hemisphere, and their names – Dorado, Tucana, Centaurus – sound romantic.

Just as the constellations that lie far to the south remain out of sight (assuming you live at a mid-northern latitude), the constellations far to the north remain visible all year. They do not change with the seasons, although they are more easily visible at one time of year than another. Because they are always available, they seem to have less value and we take them more for granted. The Big Dipper is above North America for much of the year and has little novelty – yet Australians never see it.

The most popular and best-known constellations are the 12 that make up the classical zodiac. This is true even though half of the zodiacal constellations contain few, if any, bright stars and are not conspicuous, and several cannot be seen from urban areas. They are well-known because of their astrological associations, but their fame does not correspond to their visibility. It is ironic that everyone has heard of Cancer, although few could find it, while comparatively few people have heard of Auriga, which is a magnificent constellation of bright stars that is filled with star clusters. Just as some people are famous because of where they are (the position they have in our society) rather than who they are, so it is with the constellations. A constellation's popularity tells you nothing about how interesting it is. The zodiac is treated fully in Chapter 3.3.

# History of the Constellations

Humans have the need and the skills to find patterns in nature. Our brains are programmed to invent patterns and to impose order on disorder. This talent helped our remote ancestors find their way around hunting and scavenging sites, and it helps us find our way around the sky.

We also have a need to feel connected to the cosmos. Being completely inaccessible, the sky is an endless source of mystery and wonder. It takes a soulless person indeed to gaze up at the sky on a dark, starry night and not wonder how we fit into it all. People have been doing just that since the beginning of time.

The origins of most constellations are lost in the mists of antiquity, and some of them are prehistoric. We can only unravel as much as we can of the origins and history of the oldest constellations by using the sparse clues available to us. Others were created in more recent times and their history is documented, and the far southern constellations were outlined and named during the Age of Exploration only several centuries ago. As recently as the 19th century astronomers were proposing new ones, but their number was fixed at 88 early in the 20th century and the days of creating new constellations are over.

The core of our familiar western constellations probably originated in prehistoric Sumeria. The Sumerians, who lived in the arid land between the Tigris and Euphrates Rivers in what is now southern Iraq, developed one of the world's first great civilizations and counted the invention of writing among their achievements. They were a superstitious folk, and they paid great attention to corres-

## The Oldest Constellation?

The Great Bear (the brightest part of which is the Big Dipper) is probably the oldest constellation, and it dates to prehistoric times. Bears were worshipped in "cave man" days in Europe, before they became extinct (in Europe) at the end of the last Ice Age. Bears are still worshipped by nomadic people in Siberia.

People named the celestial Great Bear for its behavior. It does not look like a bear, but it does act like a bear. Earth bears hibernate, and so does the celestial bear. It's low in the north in winter and returns in spring in a way that reminded people of bears' seasonal behavior.

What is truly interesting is that widely separated people around the world – from Europe to Asia to North America – saw these stars as a bear. We can only speculate, but it's likely that the concept of the Great Bear originated during the Ice Age and was carried from Europe to Siberia – or the other way around – and then to North America more than ten thousand years ago. If so, this association is one of the world's oldest surviving cultural artifacts.

pondences between events in the sky and events on Earth. They saw the seasons as a cyclic battle between the Lion and Bull. In mid-winter 6,000 years ago, Leo the Lion stood high overhead while Taurus the Bull lay "dying" on the western horizon, and the lion was triumphant. The reappearance of the Bull in the morning sky marked the return of spring and the death of winter, and the Bull's turn to triumph. They charted the stars to try to figure out what was happening to them and to the world around them, and their ideas of interpreting omens were elaborated upon by the Babylonians, who lived in the same area much later. The Babylonians left the first constellation lists on clay tablets, and they invented the idea of the zodiac around the 6th century BC.

Little Babylonian and less Sumerian constellation lore remains today, and that which does is imbedded in the classical Greek stories. The Greeks borrowed the concept of the zodiac from the Babylonians and incorporated Babylonian star stories into their own myths. Later, the Romans borrowed the Greek stories and we've borrowed those of the Romans, so there is a tradition of borrowing and elaborating that dates back at least 6,000 years.

We know Greek mythology in detail, but less about how they divided the sky. There are only scattered references to stars and star patterns from early Greek times.

The astronomer Ptolemy, who lived in Alexandria around 150 BC, described the 48 "Ptolemaic" or classical constellations, which remained essentially unmodified for 14 centuries. Perhaps 30 of these are Babylonian in pedigree and the rest indigenous Greek. His main source was a long poem based on an earlier and now lost work from about 350 BC by Eudoxus – the Greek astronomer who constructed the first recorded celestial globe and who worked out the idea of celestial coordinates. Ptolemy's book became known as the *Almagest* ("the Greatest") when translated into Arabic. The Arabs gave most of the stars their familiar common names, which are usually Arabic translations of the stars' positions as described by Ptolemy. Rigel, for example, comes from Arabic for "foot," which is exactly where the star is within Orion. Following the Dark Ages, the Almagest was translated into Latin, the universal language of the Christian world, and reintroduced into Europe around the year 1,000 after an absence of nearly a millennium. The 48 Ptolemaic star patterns form the core of the constellations of the northern sky.

Many additional constellations were added during the Age of Exploration. When European navigators first ventured into southern waters in the late 1500s and early 1600s, they discovered an uncharted sky in addition to uncharted lands. They divided the southern sky into groups of stars, naming the new constellations after exotic things they found in the new world, like Pavo the peacock and Indus the American Indian. Many of these new constellations achieved legitimacy and permanence by virtue of being included in Bayer's great **Uranometria** star atlas of 1603. (Bayer also introduced the idea of using Greek letters to name the brighter stars in this atlas.)

Seven additional constellations appeared in 1690 in a star atlas by the Polish astronomer Johannes Hevelius. He felt that some areas of the sky were too empty and that his atlas would look more attractive if these areas of faint stars only were filled in. His new constellations include Lacerta the Lizard and Vulpecula the Fox. Hevelius is remembered today as the last astronomer to reject using the "newfangled" telescope and to observe by eye instead.

The last burst of constellation-naming is courtesy of the French astronomer Nicolas Louis de Lacaille, who lived in Cape Town, South Africa, from 1750-1752. He created 14 new constellations

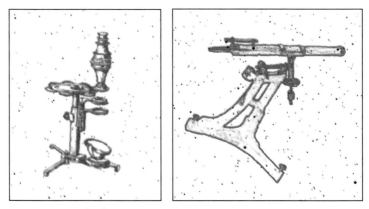

**Microscopium (left) and Telescopium, two of the southern constellations named after technological objects.**

## Defunct Constellations

Not all constellations are ancient. Although the last wave of constellation-inventing ended when the southern sky was finally filled in the middle of the 18th century, astronomers felt free to create their own as recently as the late 1800s. Often constellations were created for political purposes – to flatter a patron, for example – but such contrivances were seldom accepted graciously by other astronomers, and most disappeared as quickly as they appeared. They now are minor footnotes in the history of the sky. Examples include Robur Carolinium, or Charles' Oak, invented by Edmond Halley to honor King Charles II (who once escaped death by hiding in an oak tree); Frederick's Glory, a sword that honored Prussia's Frederick the Great; and Telescopium Herschelli or Herschel's Telescope.

from the southern stars, naming them after practical objects like an air pump, a chisel, a microscope, and a pendulum clock. This may not seem romantic to us, but apparently these were fascinating objects in their time, and now they too are immortalized in the sky. They contrast so greatly with the mythological beasts and heroes of the classical age that Lacaille has been accused of turning the sky into someone's attic.

Through the 18th century, there was no universally approved list of constellations and no official constellation boundaries. Mapmakers were free to add new constellations if they wished and to decide where one constellation ended and another began. This situation was intolerable to astronomers, who were creating ever-more detailed star charts, and an early task of the new International Astronomical Union was to settle the constellation boundary

## The Official Cygnus

Cygnus may look like a cross or even a swan, but the actual constellation includes many fainter stars that lie outside the popular stick figure. Since 1928 a constellation has been defined in the same way that a parcel of property is described on Earth – by specifying its boundaries as a series of interconnected straight lines. The true definition of Cygnus is everything in the sky that lies within these boundaries (the middle two-thirds of the description is omitted):

Méridien de 19 h. 15 m. 30 s. de 27° 30' à 30° 0'

Paralléle de 30° 0' de 19 h. 15 m. 30 s. à 19 h. 21 m. 30 s.

Méridien de 19 h. 21 m. 30 s. de 30° 0' à 36° 30' m

Paralléle de 36° 30' de 19 h. 21 m. 30 s. à 19h. 24 m.

Méridien de 19 h. 24 m. de 36° 30' à 43°

...

Paralléle de 28° 0' de 21 h. 44 m. à 21 h. 25 m.

Paralléle suite de 28° 0' de 21 h. 25 m. à 20 h. 55 m.

Méridien de 20 h. 55 m. de 28° 0' à 29° 0'

Paralléle de 29° 0' de 20 h. 55 m. à 19 h. 40 m.

Méridien de 19 h. 40 m. de 29° 0' à 27° 30'

Paralléle de 27° 30' de 19 h. 40 m. à 19 h. 15 m. 30 s.

(Reference: Délimitation Scientifique des Constellations, E. Delporte, Cambridge, 1930.)

question once and for all. A commission was appointed to draw up a list of official constellations and to define their boundaries, and the commission's results were approved in 1928. The boundaries are a series of "straight lines" in the sky that read like legal property boundaries on Earth. Starry Night will show the constellation boundaries and names. Since 1928, no new constellations have been invented.

## Constellations of Other Cultures

Our familiar constellations are a product of the history of our western culture. Today, the 88 constellations we know and love are universally recognized. Like our Gregorian calendar, they are "official" around the world. But it was not always so, and each culture invented its own way of dividing the sky. Sadly, the indigenous constellations of most cultures have been lost and only in scattered remote areas are pre-western constellations still remembered.

The ancient Egyptians saw a Crocodile, Hippopotamus, the front leg of a bull (our Big Dipper), and the god Osiris (our Orion). Their Isis is our Sirius – an important goddess whose reappearance was used to predict the annual flood of the Nile. The original Egyptian constellations were replaced by the familiar Greek constellations after Egypt was conquered by Alexander the Great in the third century BC, and in the following centuries their ancient knowledge was almost completely lost. The only clues that remain of ancient Egyptian sky lore come from enigmatic tomb paintings.

The Chinese divided the ecliptic into 28 lunar mansions, somewhat analogous to our zodiac (which is solar), and the stars into almost 300 groupings that are smaller than our constellations and that we would call asterisms. Some patterns that look so obvious to us that we have a hard time seeing them any other way, like the "W" of Cassiopeia, are subdivided by the Chinese (in the case of Cassiopeia, into three). They have four seasonal "super-constellations," each made of several asterisms with a similar theme: the White Tiger of autumn, the Black Tortoise of winter, the Blue Dragon of spring, and the Red Bird of summer. The Chinese paid less attention to star brightnesses than we do when dividing the sky and they incorporated fainter stars. We are amazed that Lynx is an official constellation, so faint are its stars, but the Chinese regularly included such faint stars in their asterisms. Among the Chinese constellations are the Dogs, the Awakening Serpent, the Wagging Tongue, the Tortoise, the Army of Yu-Lin, the Weaver Maid, and the Cow Herder.

## Peruvian Dark Constellations

The Inca of pre-conquest Peru recognized what we may call "dark constellations." At their southern latitude, the center of the Milky Way passes straight overhead and is spectacular. They recognized patterns of bright stars like we do, but they also named the dark dust clouds of the Milky Way. They thought of the dark clouds as earth carried to the sky by the celestial river. Among the dark clouds were Ya-cana the Llama and her baby Oon-yalla-macha, the fox A'-toq, the bird Yutu, and Hanp-á-tu the Toad. The Indians watched for changes in the visibility of the dark clouds, caused by upper-atmospheric moisture, that would tell them whether the coming year would be wet or dry.

# The Changing Constellations

Remember that constellations are areas of the sky, and they have infinite depth, so the stars in a constellation do not necessarily have anything to do with each other, although they appear to lie in approximately the same direction as seen from Earth. Two stars that appear to be very close to each other may actually be separated by enormous distances, one far beyond the other, while stars on opposite sides of the sky may be relatively close to each other with us in the middle. You cannot tell just by looking. This third dimension of depth means that the appearance of the constellations depends on your vantage point.

Distances within our solar system are so small that the constellations look exactly the same from Mars, Venus, and Pluto. But if we move far beyond our solar system, the story changes. If our Earth orbited a distant star rather than our Sun, the stars would be distributed in completely different patterns and our familiar constellations would not exist. If aliens live on other planets in orbit around other stars, they have their own constellations.

To see this for yourself, open the file "LDEarth" to view the stars in the Little Dipper as seen from Earth. Now open the file "LDAlpha" to view the Dipper as seen from Alpha Centauri, our nearest neighbor (about 4 light years away). Arrange the windows so that you can see both drawings of the constellations at the same time. You can see that the bowl of the Dipper has become distorted as viewed from Alpha Centauri. This distortion will increase as we move farther from our solar system, and the seven stars in the Little Dipper will make a pattern which doesn't resemble a dipper at all. Open "LDArct" to see this. You are now viewing the Little Dipper from Arcturus, which is about 40 light years away from Earth. Clear-

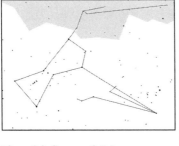

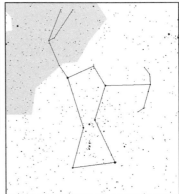

The stick figure of Orion, as seen from our Earth (right) and from Sirius.

ly the Arcturians would have a quite different name for this star grouping than the "Little Dipper"! They would probably not even group the same seven stars together. Clearly, our constellations are valid only for our solar system.

## Interstellar Exploration

In addition to all the other adjustments that come with moving into a new home, humans of the future who colonize planets around other stars will have the task of dividing their sky into their own constellations. It will be a lot of fun.

One of the nearest stars to Earth that we may visit is Barnard's Star, about six light years distant. Open the file "Barnard" to see that the new immigrants would see a bright $0^{th}$ magnitude star near the belt of Orion which is not in our night sky – because it is our own Sun!

Even when viewed from Earth, the constellations are not forever. Each star in the sky is actually in motion. Although these speeds are quite high (most stars are moving at several hundred kilometers per second!), the vast distance to the stars mean that this motion is imperceptible to us. Each star's speed and direction is different from its neighbors, causing each star to move relative to its neighbors as seen from Earth, over very long periods of time. This very slow motion of a star across the celestial sphere is called its *proper motion* ("proper" meaning "belonging to," rather than "correct").

Our familiar constellations look much the same now as they did at the end of the last Ice Age, but in the distant future they will become distorted by their stars' motions and eventually they will become unrecognizable. The constellations of 1 million AD will bear no resemblance whatsoever to the star patterns we know and love today. Eventually people will have to invent new constellations. It would be interesting to know how long people will retain the classical constellations as they become increasingly distorted before revising the scheme and starting over.

## The Big Dipper 100,000 Years from Now

The Big Dipper really does look like a dipper, but it won't always. The middle five stars are traveling together through space on nearly parallel paths as they orbit around the center of the Milky Way, and they will retain their relative spacing far into the future. The Dipper's two end stars, however, are traveling in the opposite direction. Eventually the Big Dipper will become stretched into what people may one day call the Big Lounge Chair.

# HIGHLIGHTS OF THE CONSTELLATIONS

Starry Night shows all 88 constellations in the sky. Many constellations – especially the newest ones such as Lynx – are obscure and are of little interest. They have no bright stars and it can be a challenge simply to identify them, even on a dark night away from city lights. Some are eternally below your southern horizon (or northern horizon, if you live in the Southern Hemisphere) and permanently out of sight. But the major constellations are signposts in the sky and they contain well-known bright stars, star clusters, and nebulae that are within the reach of binoculars or small telescopes. A few minor constellations also contain objects of special interest. See Chapter 5.1 for descriptions of the different kinds of star clusters, nebulae, galaxies, double stars, and variable stars.

This chapter describes a sampling of bright stars and well-known deep-space objects within selected constellations. The constellations are organized according to the Northern Hemisphere season when they are best viewed during the evening. See Appendix A for a list of all 88 constellations.

There is a file for each constellation described in this chapter. To view the files for this chapter in the best manner, when you open Starry Night, select "Constellations | Constellation Settings" and make sure that the boxes marked "Name" and "Stick figure" are checked.

This will ensure that the constellation files that you open show the boundary and stick figure of the relevant constellation.

# Winter

*ORION* (THE HUNTER) dominates the winter sky. Three bright stars in a row form his belt, two stars mark his shoulders, and two stars mark his knees or feet. A sword hangs from his belt.

The two brightest stars in Orion are Betelgeuse and Rigel. They lie at opposite ends of Orion – Betelgeuse is in his left shoulder and Rigel in his right foot – and they have contrasting colors. Betelgeuse is known for its orangeness, while Rigel is the bluest bright star in the sky. Rigel is also a pretty double star.

Betelgeuse is enormous. If our Sun were placed at its center, the planets out to and including Mars would orbit within it. Such supergiant stars are unstable, and Betelgeuse varies its size and its brightness slightly and irregularly as it expels material into space. It is 10,000 times as luminous as our Sun and lies at a distance of about 520 light years. Betelgeuse is well on the road to exploding as a supernova.

Rigel is almost twice as distant at 900 light years, and is also more luminous (50,000 suns). It is approximately the same size as Betelgeuse. It is a blue supergiant star with a surface temperature twice that of our Sun. It has a 7th-magnitude companion star 9 arcseconds distant (their true separation is about 50 times the distance from Pluto to the Sun).

Mintaka (also known as Delta Orionis) is the westernmost of the three bright stars that form Orion's belt. It is one of the prettiest double stars in the sky for a small telescope. The main star is magnitude 2.2, and its 5.8-magnitude companion lies 53 arcseconds distant. At a distance of 1,600 light years their true separation is about 1/2 light year, or 30,000 times the distance between the Earth and Sun.

The most famous object in Orion is the famous Orion Nebula (M42) – a true showpiece in a telescope of any size. It is the fuzzy naked-eye "star" that marks the jewel in Orion's sword. A telescope reveals it to be a glowing cloud of hot hydrogen gas, illuminated by a group of hot stars (the Trapezium) at its center that are causing the nebula to fluoresce. The

The Orion Nebula.

visible part of the nebula is 30 light years across and larger in apparent extent than the full Moon. It lies about 1600 light years from Earth. A detached portion to the south (separated by a dark lane of dust that lies in the foreground) is called M43. The Orion Nebula's outer parts extend over 5° from M42 and cover much of Orion.

Open the file "Orion" to see Orion with major stars labeled and the Orion Nebula.

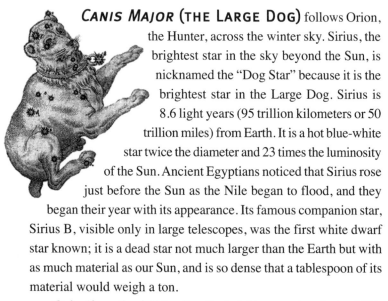

**CANIS MAJOR (THE LARGE DOG)** follows Orion, the Hunter, across the winter sky. Sirius, the brightest star in the sky beyond the Sun, is nicknamed the "Dog Star" because it is the brightest star in the Large Dog. Sirius is 8.6 light years (95 trillion kilometers or 50 trillion miles) from Earth. It is a hot blue-white star twice the diameter and 23 times the luminosity of the Sun. Ancient Egyptians noticed that Sirius rose just before the Sun as the Nile began to flood, and they began their year with its appearance. Its famous companion star, Sirius B, visible only in large telescopes, was the first white dwarf star known; it is a dead star not much larger than the Earth but with as much material as our Sun, and is so dense that a tablespoon of its material would weigh a ton.

Only 4° south of Sirius lies the bright open star cluster M41. Its 80 stars, the brightest of which are 7th magnitude, fit within a circle 2/3° in diameter. M41 is 25 light years in diameter and 2,000 light years distant. Because of its large size, it is prettiest in binoculars or a low-power eyepiece.

Open the file "Canis Major" to see Canis Major, Sirius, and M41.

**TAURUS (THE BULL)** is an ancient constellation that dates back to when bulls were worshipped in the Middle East. We see it as a face with a red eye and two menacing horns.

Aldebaran – the eye of the bull – is an orange giant star about 150 times as luminous as our Sun. It is one of three bright orange stars in the sky, the other two being Betelgeuse in Orion and Antares in Scorpius. It appears amid the stars of the Hyades Star Cluster, but with a distance of 68 light years from Earth it is actually well in front of and only half as distant as the cluster's stars.

The face of the bull is outlined by the remarkable Hyades Star Cluster – the closest bright star cluster to Earth. Recent measurements reveal that its center is 151 light years from Earth and it has a diameter of 15 light years. A handful of its stars are visible to the unaided eye, dozens can be seen with binoculars, but the cluster's membership totals 200 known stars. The part we see by eye is 5-1/2° in diameter, which corresponds to 15 light years, but this is only its core; fainter stars lie up to 12° (40 light years) from its center. This is an old star cluster with an estimated age of 700 million years, and many of its stars are yellow – in contrast to the youthful Pleiades with its blue stars.

The prettiest star cluster in the entire sky is certainly the Pleiades, also known as the Seven Sisters. The Pleiades rises an hour before the main part of Taurus, and in Ptolemy's time it was considered a separate constellation. A minor

The Pleiades star cluster.

mystery is why it should be called the "Seven Sisters" when only six stars are brighter than the rest, but probably the facts were ignored to fit the mythology of the sisters, who were seven daughters of Atlas. Some people can see 10 or 11 stars, binoculars reveal dozens, and the total count is probably over 500. These are young stars born only about 100 million years ago, and the cluster is not much older than the Rocky Mountains. Blue stars, which burn out relatively quickly, remain, and they illuminate with bluish light a dust cloud that is passing by (visible only in long-exposure photos). The Pleiades is 380 light years from Earth and about 12 light years in diameter.

Open the file "Taurus" to see Taurus and to see Aldebaran, the Hyades, and the Pleiades.

# Spring

***GEMINI* (THE TWINS)** are two bright stars – Castor and Pollux – and two strings of fainter stars that extend from them and that can be made to resemble two brothers standing side-by-side. Their exploits were legendary in the classical world.

Castor is a sextuple star (two, which are bluish, are visible through a telescope) that lies 52 light years from Earth. Yellowish Pollux lies 18 light years nearer to Earth than Castor and the two are physically unrelated.

Near Castor's foot is the bright 5th-magnitude open star cluster M35. It is as large as the full Moon and is easily visible to the un-aided eye on a dark night, and it is a pretty sight in binoculars. The cluster, which contains at least 200 stars, lies 2,700 light years from Earth and is 20 light years in diameter.

Open the file "Gemini" in the "Constellations" folder to see Gemini and to see Castor, Pollux, and M35.

***CANCER* (THE CRAB)** is a small constellation with faint stars only, and it cannot be seen from within a city. Most people have heard of it only because the Sun, Moon, and planets pass through it.

Cancer's major feature of interest is the Beehive Star Cluster (also called the Praesepe – Latin for "manger"). At 3rd magnitude, it can easily be seen without binoculars, and with a diameter of more than 1° it is too large to fit within the field of view of most telescope eyepieces. The Beehive contains at least 50 stars, is 13 light years in diameter, and is 580 light years from Earth. The Beehive was one of four star clusters known in antiquity (the others being the Hyades, Pleiades, and Coma Berenices).

Open "Cancer" to see Cancer and the Beehive Star Cluster. Zoom in to see the Beehive's brightest stars.

M44, the Beehive Cluster.

**COMA BERENICES (BERENICE'S HAIR)** was named in honor of Queen Berenice of ancient Egypt. It lies one-third of the way between the tail of Leo and the end of the handle of the Big Dipper.

Coma Berenices is both a constellation and a star cluster. To the unaided eye it is an indistinct cluster of faint stars almost five degrees across – almost as large as the bowl of the Big Dipper. Binoculars or a telescope reveals several dozen stars 290 light years distant.

Open the file "Coma Berenices" to see the constellation; zoom in to see the individual stars that make up the open cluster (it is in the bottom right corner, just beneath Gamma Comae Berenices, which is the last star in the stick figure). Note how the distances of these stars are all around 290 or 300 light years - if a star's distance is much different from this, it is not part of the cluster.

**URSA MAJOR (THE GREAT BEAR)** is a huge constellation (the third largest) that lies far to the north and is best seen in the

Northern Hemisphere's spring and summer. The most famous part of it are the seven brightest stars that form the Big Dipper (as it is called in the United States and Canada), but the rest of the bear extends to the west and south.

## Big Dipper

The Big Dipper is so obvious and well known that you might think it has always been called that – but not so. It is called a dipper only in the United States and

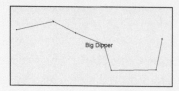

Canada, and there it has been called a dipper for only about 200 years. Its origin is lost in time, but the idea of a dipper might have come with slaves from Africa, where they drank from hollowed gourds. Before water came from taps, people used dippers to scoop water out of buckets.

In England the Big Dipper is called the Plough, and in Germany it is the Wagon.

Mizar is the second star from the end of the handle of the Big Dipper, and is actually part of a multiple star system, as we saw in Chapter 5.1. People with better-than-average eyesight will see a slightly fainter star, Alcor, 12 arcminutes distant. The two names come from "horse" and "rider" in old Arabic, and according to legend they were used as an eye test. Mizar is 78 light years distant and Alcor lies 3 light years beyond.

Mizar itself is a very pretty double star in a small telescope. It has a 4th-magnitude companion star that lies 14 arcseconds distant and that is an easy target for even the smallest telescopes. Their true separation is at least 10 times the distance from the Sun to Pluto. Mizar was the first double star discovered, in 1650.

Open the file "Ursa Major" in the "Constellations" folder to see the Big Dipper and Mizar and Alcor.

# Summer

*CYGNUS* (THE SWAN) is a graceful bird with outstretched wings in classical mythology. His head is the star Albireo, his tail the bright white star Deneb, and his wings reach symmetrically to the side. Cygnus was renamed the "Cross" by Schiller in his unsuccessful attempt to Christianize the constellations, and the name stuck – although it is unofficial. It is often called the Northern Cross (equally unofficially) to distinguish it from the Southern Cross.

Cygnus straddles the Milky Way in the summertime. Use binoculars or a low-power telescope to see the myriad of faint stars within it. The Milky Way is especially rich in this direction because when we look toward Cygnus, we are looking down the length of one of the spiral arms. The dark Great Rift – a split in the Milky Way that begins north of the middle of Cygnus and that continues far to the south – is the effect of enormous dark clouds of gas and dust that block the light of millions of faint stars that lie beyond.

Two especially noteworthy stars are Deneb and Albireo. Deneb, the brightest star in the constellation and one of the stars of the Summer Triangle, is a true supergiant. Its blue-white color distinguishes it from other giants which are orange, such as Antares and Betelgeuse. Deneb is 50,000 times as luminous as our Sun. It lies at the enormous distance of 1,600 light years – so far away that the light we see tonight left Deneb at the end of the Roman Empire. Albireo, which marks the Swan's head, is a pretty double star in a small telescope with contrasting colors (often described as blue and gold) separated by a comfortable 34 arcseconds. They are giant stars 400 light years from Earth and at least 650 billion kilometers (100 times the distance from Pluto to the Sun) from each other.

Open the file "Cygnus" to see the stick figure with bright stars highlighted, and the Milky Way.

*HERCULES* (THE STRONG MAN) is the ancestor of Superman and other superheroes who have extraordinary strength and skill. He was formerly called the Kneeler for reasons long lost. Hercules is a rough box of medium-bright stars – the Keystone – plus others nearby that can be made to look like a man only with great difficulty.

Hercules contains the famous M13 Globular Star Cluster, one of the showpieces of summer star parties since it passes nearly overhead when seen from North America or Europe, and is easy to find. Look for it 1/3 of the way from Eta to Zeta Herculis on the western segment of the Keystone. At 6th magnitude it is barely visible to the naked eye under ideal conditions, and through a small telescope it looks like a small fuzzy patch of light. A large telescope reveals that it is made of thousands of stars, the brightest of which are 12th magnitude. The cluster is about 23,000 light years distant and about 150 light years in diameter.

**M13, the Hercules Cluster.**

Open the file "Hercules" to see Hercules and the globular cluster M13.

*LYRA* (THE LYRE) is a small constellation whose brightest star, Vega, is the westernmost of the three stars of the Southern Triangle. A lyre is an ancient stringed instrument which is the ancestor of the harp and the guitar. Lyra is overhead during the summer in North America.

Vega is the fifth brightest star in the sky. It is also relatively nearby at a distance of 25 light years (55 trillion kilometers or 35 trillion miles). It is a large white star about 50 times as luminous and 2-1/2 times as massive as our Sun. Precession of the equinoxes will cause it to become the Pole Star in 14,000 AD. Vega is consuming its fuel rapidly and will burn out in less than one billion years.

Near Vega is the famous double-double star Epsilon Lyrae. People with near-perfect eyesight can see that this 5th-magnitude star is actually two "stars" separated by a wide 209 arcseconds. Under magnification, each star is revealed to be a pair of stars separated by a close 2-1/2 arcseconds. Each star is approximately 100 times the luminosity of our Sun, and they are about 130 light years from Earth.

The famous Ring Nebula (M57) is the best known "planetary" nebula (see Chapter 5.1) – a tenuous shell of hot gas expelled from a dying star. Through a telescope it looks like a tiny slightly oval donut just over 1 arcminute in diameter. Its true diameter is 1/2 light years and its distance is over 1,000 light years.

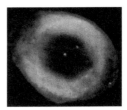

The Ring Nebula, M57.

Open the file "Lyra" to see Lyra, the stars Vega and Epsilon Lyrae, and the Ring Nebula.

*SAGITTARIUS* (THE ARCHER) is a centaur, according to mythology that dates to Sumerian times. Centaurs, who were half-man and half-horse, combined the skill of men with the speed of horses and were not to be messed with. Today Sagittarius is easily seen as the outline of a teapot. It lies in the southern sky during summer.

The constellation lies in front of an exceptionally rich part of the Milky Way, whose center lies within its boundaries, and it contains a wealth of objects for amateurs with telescopes. Only four are described here.

The Lagoon Nebula (M8) is a 5th-magnitude glowing cloud of hot gas larger than the full Moon and bright enough to see with the naked eye on a dark summer night. Only one other nebula (the Orion Nebula, which is visible in the opposite season) is more spectacular through a telescope. It fluoresces due to the presence of hot young stars near it that recently formed out of its gaseous material; a cluster of new stars nearby is called NGC 6530. The nebula is about 5,000 light years distant and about 50 by 100 light years in dimension.

The Omega Nebula (M17) is so named because its arclike shape resembles the capital Greek letter Omega. Also called the Swan or Horseshoe, this fluorescing diffuse nebula is bright enough to be visible in binoculars. It is about 6,000 light years distant and 40 light years across.

**The Omega Nebula.**

The Trifid Nebula (M20) is smaller than the nearby Lagoon, but it is easier to see in a telescope because it has a higher surface brightness. Its name comes from a complex dust lane that lies in front of and trisects it. The Trifid is just 1-1/2° from M8 and the two can be seen together in a spotting scope or binoculars. The Trifid lies 6,500 light years from Earth and is about 30 light years in diameter.

M22 was the first globular cluster discovered, in 1665. It is visible to the naked eye on a dark night, and through a telescope it rivals M13 in Hercules. M22 would be more magnificent than M13 for observers in the United States and Canada were it higher in the sky, but its southern declination puts it at a disadvantage. M22 is one of the closest globulars, with a distance of 10,500 light years from Earth, and its brightest stars are 11th magnitude – a full magnitude brighter than the stars of M13.

M22 is one of several bright globular clusters in Sagittarius (others are M28, M54, M55, M69, M70, and M75). Early last century the astronomer Harlow Shapley proposed that globular clusters are concentrated towards Sagittarius because they orbit the center of the Milky Way, and they all pass through this part of the sky before dispersing to the outer parts of their orbits. He concluded that the Milky Way's center must lie in the direction of Sagittarius and estimated the distance to it, displacing the Earth from its apparent position at the center of our galaxy.

Open the file "Sagittarius" to see the constellation and to see the globular clusters and nebulae within it.

## The Closest Galaxy

The closest galaxy to the Milky Way is the Sagittarius Dwarf Elliptical Galaxy (SagDEG) which is half as distant as the two Magellanic Clouds. It circles the Milky Way with an orbital period of about one billion years. The 8th-magnitude globular star cluster M54, long believed to lie within the Milky Way, was found in 1994 to be at the enormous distance of 80,000 light years. This puts it outside the limits of our Milky Way but inside the SagDEG. It is the most distant globular star cluster visible in all but the very largest amateur telescopes and intrinsically the most luminous. Its brightest stars are 14th magnitude.

M54, a star cluster in the nearest galaxy to our own.

*SCORPIUS* (THE SCORPION) has been called a scorpion for at least the last 6,000 years, and its stars do indeed look like a scorpion. Too far to the south to see well from Canada, it is prominent in the south in the summer sky from the southern United States. At one time its claws included the stars of Libra, to the west.

The Scorpion's heart is a bright red star (actually orange) called Antares – the "rival of Mars" in Greek because it rivals Mars in color and is often close to Mars in the sky (see the sidebar below). Antares is a true supergiant star 10,000 times as luminous as our Sun and so large that if it were placed where our Sun is, Mars would orbit beneath its surface. It lies about 500 light years from Earth – the same distance as the orange star Betelgeuse in Orion. Like all giant stars, it varies slightly in brightness in an irregular pattern and will soon – astronomically speaking – burn out.

Beta Scorpii (also known as Graffias) is an exceptionally pretty double star. The two stars are magnitudes 2.6 and 4.9 with a separation of 14 arcseconds, and it closely resembles Mizar in the

## Mars and Anti-Mars Together

Antares lies near the ecliptic at an ecliptic latitude of -4° 34', and the Sun, Moon, and planets pass near it. Mars is in conjunction with its rival Antares on the average of once every two years. Mars-Antares conjunctions occur in March 2001, January 2003, January 2005, December 2006, November 2008, November 2010, and so on. Open the file "Antares" and step through time to see the January 2005 conjunction. If you wish, you can continue to run time forward or backward to see other conjunctions.

Big Dipper. The pair is 600 light years distant.

The brightest globular star cluster in Scorpius is M4. It is magnitude 5-1/2 and it lies only 1-1/3° west of Antares, so it is very easy to find. This exceptionally close (for a globular!) star cluster is 7,000 light years distant and 55 light years in diameter. Its brightest stars, like M22's in Sagittarius, are 11th magnitude.

Two especially bright open star clusters are M6 and M7, which lie above the stinger of the Scorpion's tail. At magnitudes 5 and 4 respectively, they are easily visible to the unaided eye if Scorpius is high enough in the sky from your latitude, and both can be seen together in binoculars. They are 2,400 and 800 light years distant and each contains about 80 stars. Ptolemy called M7 a "nebula," and its identity as a star cluster was not known until the invention of the telescope. Because these clusters have very large angular sizes, observe them with very low magnification.

Open the file "Scorpius" to see Scorpius and to see Antares, Beta Scorpii, and the star clusters M4, M6, and M7.

# Autumn

***ANDROMEDA* (THE PRINCESS)** is a line of three bright stars and a string of fainter stars that forms a long narrow V in the autumn sky. Alpha Andromedae is one of the four stars that forms the Square of Pegasus.

Andromeda contains one of the prettiest double stars in the sky – Gamma Andromedae, also called Almach. The two stars, which are magnitudes 2.2 and 4.9, are noted for their contrasting colors – yellow and bluish respectively. This color contrast is shown well with Starry Night. These are giants 90 and

15 times the luminosity of our Sun. The angular separation between them is 10 arcseconds (1/6 of 1/60 of one degree) and their distance from Earth is 350 light years.

To cosmologists, the constellation Andromeda is almost synonymous with the Andromeda Galaxy – the closest and brightest large galaxy to us. It is visible to the unaided eye on a dark night as a small fuzzy patch of light, but even with a large telescope it remains a disappointingly shapeless blob. We see only the inner part of what time-exposure photographs reveal to be a huge spiral galaxy of hundred of billions of stars that spans several degrees of the sky. It is similar in size and shape to the Milky Way; our Milky Way would look something like the Andromeda Galaxy from a similar distance. At a distance of almost 3 million light years, it is the most distant object

**M31, the Andromeda Galaxy.**

you can see without a telescope. It has two satellite elliptical galaxies, M32 and NGC 205, which are easily visible in small telescopes. The Andromeda Galaxy is also known as M31 – the 31st object in Charles Messier's catalog of cometlike objects.

Open the file "Andromeda" to see the constellation outline and stick figure with its major stars and M31 highlighted.

**CEPHEUS (THE KING)** is an inconspicuous five-sided figure next to Cassiopeia in the far northern sky. From the northern United States it is circumpolar, and it never sets.

Although just a faint star to the unaided eye, Delta Cephei is one of the most important and most studied stars of all. It is the prototype of the famous

Cepheid Variable Stars – unstable giant stars that pulsate in size and that vary in brightness in a regular way. The time it takes a Cepheid star to go from its maximum brightness down to its minimum brightness and back to its maximum again is called its *period*. The intrinsic brightness of these stars is related to the lengths of their periods, and this makes them a powerful tool for finding distances to galaxies. Because there is a one-to-one relationship between the period of variability and the brightness of a Cepheid variable, once its period is measured its intrinsic brightness is known, and since we can observe how bright the Cepheid **appears** to be, we can calculate its actual distance. The largest telescopes can see Cepheid Variables in galaxies 50 million light years distant.

Delta Cephei is approximately 3,000 times more luminous than our Sun and it lies some 1000 light years from Earth. It varies in brightness by a factor of two from magnitude 3.6 to 4.3 in a period of precisely 5 days, 8 hours, and 48 minutes. Compare it to nearby 4.2-magnitude Epsilon Cephei and 3.4-magnitude Zeta Cephei to monitor its regular change in brightness from your backyard.

Open the file "Cepheus" to see Cepheus in stick figure outline, and Delta Cephei.

**PERSEUS (THE HERO)** is a scraggly stream of stars near Cassiopeia and Andromeda in the autumn sky.

Algol is the "demon star," its name coming from Arabic for "the ghoul." By the time of the ancient Greeks, observers were alarmed that the star brightens and fades in a regular pattern. Today we know that the star is actually two stars in orbit around a common center of gravity and so aligned that they eclipse each other as seen from Earth. Normally they appear as a single star of magnitude 2.1, but when the brighter star is hidden by the fainter, it fades to magnitude 3.4 for about 10 hours. The entire cycle lasts 2 days, 20 hours, and 49 minutes. Compare with the nearby star Gamma Andromedae, which remains at a constant magnitude 2.1, and Epsilon Persei, which is magnitude 2.9. Algol is about 100 light years distant. (See the section on variable stars in Chapter 5.1.)

The Perseus Double Star Cluster is one of the more striking naked-eye deep-space objects in the sky. Visible to the eye as two "out-of-focus" stars that lie near to each other in the Milky Way, binoculars reveal them to be two giant but distant clusters of hundreds of stars. Their centers are 1/2° apart, but their edges overlap. The clusters are each magnitude 4-1/2, and their brightest stars are 6th magnitude. At a distance of about 7,200 light years, these stars are true giants with 100,000 times the luminosity of our Sun. The clusters lie in the Perseus Arm of the Milky Way – the next spiral arm beyond the Orion Arm.

Open the file "Perseus" to see Perseus and Algol. The Perseus double cluster is not included in Starry Night.

# The Southern Sky

As we learned in Chapter 3.2, the equator is the only place on Earth where all stars are above the horizon at some time. The farther north or south that we move from the equator, the more stars are forever hidden to us. The constellations described in this section are located south of the celestial equator and cannot be seen well (if at all) in the north.

*CARINA* (THE KEEL) is part of the classical Ptolemaic constellation Argo, the ship. Lacaille subdivided this enormous ancient constellation into three, of which Carina is the southernmost part.

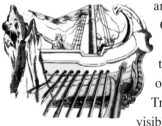

Canopus, the second brightest star in the sky (after Sirius), is named after the pilot of the fleet of Spartan ships that sailed to Troy to recover Queen Helen. It is barely visible from the southernmost parts of the United States. Canopus is a yellowish supergiant star with a diameter 65 times that of our Sun, a luminosity of 15,000 suns, and a distance of 310 light years from Earth.

The giant and explosively variable star Eta Carinae is surrounded by a nebula where stars are now forming. The Eta Carinae Nebula – the largest in the sky – is visible to the naked eye as a huge splotch of light 2° across. It is over 9,000 light years from Earth. A dark cloud in front of it is called the Keyhole Nebula.

Open the file "Carina" to see Canopus and the star Eta Carinae. Although an image of the Eta Carinae nebula is not included in Starry Night, you can select the star Eta Carinae and then choose "Edit | Online Info" to get information and images of this nebula.

*CENTAURUS* (THE CENTAUR) lies well below the celestial equator. Its northern part can be seen from the southern States, but its southernmost stars cannot be seen from the United States (other than Hawaii). The closest star to Earth, other than the Sun, is the triple star system collectively known as Alpha Centauri. It lies 4.3 light years from Earth, or 300,000 times as far as our Sun. Alpha Centauri A (also known as Rigil

Kentaurus) has a bright companion (Alpha Centauri B), which orbits in an 80-year period; they presently are 14 arcseconds apart and a magnificent pair in a small telescope. Another companion, Proxima Centauri, is a faint red dwarf star trillions of light years distant from the main two and slightly closer to Earth than the main pair. Beta Centauri (also known as Hadar) lies less than 5° to the west of Alpha, and the two are a striking pair of very bright stars near the Southern Cross.

The 4th-magnitude Omega Centauri Globular Star Cluster is the biggest and most luminous globular star cluster in the Milky Way (it has perhaps a million stars) and also the brightest in our sky. Its nature was discovered in 1677 by Edmund Halley (of Halley's Comet fame); earlier it was mistakenly considered a star, which is why it has a Bayer letter. To the naked eye it is a fuzzy ball of light, but through a small telescope it is a stunning sight. It lies approximately 50,000 light years from Earth.

 Open the file "Centaurus" to see Centaurus and Alpha and Beta Centauri. Omega Centauri is not included in Starry Night.

**CRUX (THE SOUTHERN CROSS)** is the smallest constellation. The compact group of five bright stars is the best-known star pattern in the southern sky. It guides you to where  the South Star would be if there were one: follow the long arm of the cross from Gamma through Alpha and continue it 4-1/2 cross-lengths (27°) southward to find the South Celestial Pole. The Southern Cross is a familiar icon in Australia and New Zealand, where you will see it on every-thing from flags and stamps to beer. It is visible from Hawaii and points south.

The Jewel Box Open Star Cluster (also knows as the Kappa Crucis Cluster) is a large 4th-magnitude open star cluster that is easily visible to the naked eye and very pretty in binoculars. The

Jewel Box is 20 light years across and about 7000 light years distant, and it contains at least 200 stars, the brightest of which are 7th magnitude.

One of the most unusual objects in Crux is what looks like a hole in the Milky Way that is about 5° in diameter. The Coal Sack Dark Nebula is a huge cloud of dark dust, 20 to 30 light years across and perhaps 600 light years distant, that blocks the light of stars beyond. It is the most conspicuous of the many dark clouds that lie along, and partially obscure, the Milky Way.

Open the file "Crux" to see the Southern Cross, the Coal Sack, and the Jewel Box. To find the Jewel Box, zoom in on the star labeled "Mimosa." Then zoom in to a field of view of about 5°. The Jewel Box is visible as a patch of stars about 1.5° directly above Mimosa (look for the star Kappa Crucis). The Coal Sack is the region with very few stars that is up and to the left of the Jewel Box.

This concludes our brief tour through the constellations, but it barely scratches the surface. The night sky is filled with uncountable wonders, and with a telescope you could explore the constellations forever.

# CONCLUSION

The sky is a fascinating place – especially if you understand it. It is so removed from our normal experiences that celestial objects and their slow movements can seem too abstract to relate to. Some people are overwhelmed and conclude that understanding the sky is "beyond them"; others (like you) put in the time and trouble to puzzle it all out. You have a lot of help in the form of this book and software package.

When I was a youngster, I taught myself the constellations through a few books and a rotating star finder, and I learned how objects in the sky move by reading about them and by studying diagrams in books. I had to do a lot of mental conversions to connect words, static illustrations, and tables of figures with what I saw in the sky, but in time I figured it out. It was not until desktop astronomy software came along many years later that I was really able to visualize, for the first time, what these movements looked like. I had thought about the retrograde loop of Mars and had followed it through the months – but now I could see it in minutes. I had thought about why eclipses did not happen each month – but now I could see how the paths of the Sun and Moon intersect and how that intersection point moves from year to year. I had thought about how precession causes the vernal equinox to shift – but now I could skip through centuries and watch the equinox precess. I had thought about how the planets would appear to orbit as seen from the Sun – but now I could put myself at the Sun and watch them move. To one who learned it the hard way, this was magical.

You, dear reader, are fortunate to have a tool such as Starry Night to help you understand the sky. Work with it, play with it, fool around with it, and gain wisdom.

The reward is a feeling of being connected with the sky. Good luck, and may you never stop learning about the sky. May you be at home under it wherever you go.

– John Mosley

# APPENDIX A
# THE CONSTELLATIONS

| Common Name | Latin Name | Abb. | Possessive |
|---|---|---|---|
| Princess | Andromeda | And | Andromeda |
| Air Pump | Antlia | Ant | Antliae |
| Bird of Paradise | Apus | Aps | Apodis |
| Water Carrier | Aquarius | Aqr | Aquarii |
| Eagle | Aquila | Aql | Aquilae |
| Altar | Ara | Ara | Arae |
| Ram | Aries | Ari | Arietis |
| Charioteer | Auriga | Aug | Aurigae |
| Herdsman | Boötes | Boo | Bootis |
| Chisel | Caelum | Cae | Caeli |
| Giraffe | Camelopardalis | Cam | Camelopardalis |
| Crab | Cancer | Cnc | Cancri |
| Hunting Dogs | Canes Venatici | CVn | Canum Venaticorum |

| Common Name | Latin Name | Abb. | Possessive |
|---|---|---|---|
| Large Dog | Canis Major | CMa | Canis Majoris |
| Small Dog | Canis Minor | CMi | Canis Minoris |
| Sea Goat | Capricornus | Cap | Capricorni |
| Keel | Carina | Car | Carinae |
| Queen | Cassiopeia | Cas | Cassiopeiae |
| Centaur | Centaurus | Cen | Centauri |
| King | Cepheus | Cep | Cephei |
| Sea Monster or Whale | Cetus | Cet | Ceti |
| Chameleon | Chamaeleon | Cha | Chamaeleontis |
| Compasses | Circinus | Cir | Circini |
| Dove | Columba | Col | Columbae |
| Berenice's Hair | Coma Berenices | Com | Comae Berenices |
| Southern Crown | Corona Australis | CrA | Coronae Australis |
| Northern Crown | Corona Borealis | CrB | Coronae Borealis |
| Crow | Corvus | Crv | Corvi |
| Cup | Crater | Cra | Crateris |
| Southern Cross | Crux | Cru | Crucis |
| Swan | Cygnus | Cyg | Cygni |
| Dolphin | Delphinus | Del | Delphini |
| Swordfish | Dorado | Dor | Doradus |
| Dragon | Draco | Dra | Draconis |
| Colt | Equuleus | Equ | Equulei |
| River | Eridanus | Eri | Eridani |
| Furnace | Fornax | For | Fornacis |
| Twins | Gemini | Gem | Gemini |
| Crane | Grus | Gru | Gruis |

| Common Name | Latin Name | Abb. | Possessive |
|---|---|---|---|
| Strong Man | Hercules | Her | Herculis |
| Clock | Horologium | Hor | Horologii |
| Water Serpent (f) | Hydra | Hya | Hydrae |
| Water Serpent (m) | Hydrus | Hyi | Hydri |
| Indian | Indus | Ind | Indi |
| Lizard | Lacerta | Lac | Lacertae |
| Lion | Leo | Leo | Leonis |
| Small Lion | Leo Minor | LMi | Leonis Minoris |
| Hare | Lepus | Lep | Leporis |
| Scales | Libra | Lib | Librae |
| Wolf | Lupus | Lup | Lupi |
| Lynx | Lynx | Lyn | Lyncis |
| Lyre | Lyra | Lyr | Lyrae |
| Table Mountain | Mensa | Men | Mensae |
| Microscope | Microscopium | Mic | Microscopii |
| Unicorn | Monoceros | Mon | Monocerotis |
| Fly | Musca | Mus | Muscae |
| Carpenter's Square | Norma | Nor | Normae |
| Octant | Octans | Oct | Octantis |
| Serpent Bearer | Ophiuchus | Oph | Ophiuchi |
| Hunter | Orion | Ori | Orionis |
| Peacock | Pavo | Pav | Pavonis |
| Flying Horse | Pegasus | Peg | Pegasi |
| Hero | Perseus | Per | Persei |
| Fire Bird | Phoenix | Phe | Phoenicis |
| Painter's Easel | Pictor | Pic | Pictoris |
| Fishes | Pisces | Psc | Piscium |

| Common Name | Latin Name | Abb. | Possessive |
|---|---|---|---|
| Southern Fish | Piscis Austrinus | PsA | Piscis Austrini |
| Stern | Puppis | Pup | Puppis |
| Mariner's Compass | Pyxis | Pyx | Pyxidis |
| Reticle | Reticulum | Ret | Reticuli |
| Arrow | Sagitta | Sge | Sagittae |
| Archer | Sagittarius | Sgr | Sagittarii |
| Scorpion | Scorpius | Sco | Scorpii |
| Sculptor's Apparatus | Sculptor | Scl | Sculptoris |
| Shield | Scutum | Sct | Scuti |
| Serpent | Serpens | Ser | Serpentis |
| Sextant | Sextans | Sex | Sextantis |
| Bull | Taurus | Tau | Tauri |
| Telescope | Telescopium | Tel | Telescopii |
| Triangle | Triangulum | Tri | Trianguli |
| Southern Triangle | Triangulum Australe | TrA | Trianguli Australis |
| Toucan | Tucana | Tuc | Tucanae |
| Great Bear | Ursa Major | UMa | Ursae Majoris |
| Small Bear | Ursa Minor | UMi | Ursae Minoris |
| Sails | Vela | Vel | Velorum |
| Virgin | Virgo | Vir | Virginis |
| Flying Fish | Volans | Vol | Volantis |
| Fox | Vulpecula | Vul | Vulpeculae |

The "Common Name" is often the English equivalent of the Latin name, but it can be a descriptive term too (as in the "Twins" for Gemini). "Abb." is the official three-letter abbreviation. The possessive "of" form is used following a star's name, as in Alpha Centauri – "the Alpha star of Centaurus." Those who studied Latin might recall that the possessive form is the Latin genitive case.

# APPENDIX B
# PROPERTIES OF THE PLANETS

| Planet | Mean Distance From Sun (AU*) | Radius (Earth radii) | Mass (Earth masses) | Length of Year (Earth years) | Length of Day (Earth solar days) | Inclination of Orbit to the Ecliptic Plane (°) |
|---|---|---|---|---|---|---|
| Mercury | 0.39 | 0.38 | 0.06 | 0.24 | 175.94 | 7 |
| Venus | 0.72 | 0.95 | 0.82 | 0.62 | 117 | 3.4 |
| Earth | 1 | 1 | 1 | 1 | 1 | 0 |
| Mars | 1.52 | 0.61 | 0.11 | 1.88 | 1.03 | 1.9 |
| Jupiter | 5.20 | 11.21 | 317.83 | 11.86 | 0.41 | 1.3 |
| Saturn | 9.55 | 9.50 | 95.16 | 29.42 | 0.44 | 2.5 |
| Uranus | 19.22 | 8.01 | 14.50 | 83.75 | 0.72 | 0.8 |
| Neptune | 30.11 | 7.77 | 17.20 | 163.7 | 0.67 | 1.8 |
| Pluto** | 39.54 | 0.36 | 0.002 | 248.0 | 6.39 | 17.1 |

\* 1 AU is the average Earth-Sun distance, which is 149,600,000 km (92,900,000 miles).

\*\* Pluto is sometimes no longer considered a planet.

# GLOSSARY

**absolute magnitude:** the apparent brightness (magnitude) a star would have if it were 32.6 light years (10 parsecs) from the Earth. It is used to compare the true, intrinsic brightnesses of stars. The Sun has an absolute magnitude of +4.8.

**annular eclipse:** solar eclipse where the Moon passes directly in front of the Sun, but is too far from the Earth to completely cover the Sun. A ring of sunlight surrounds the Moon at the peak of the eclipse.

**Antarctic Circle:** the line of latitude on the Earth's surface that is 23-1/2° north of the South Pole. The Antarctic Circle marks the northernmost points in the Southern Hemishpere that experience the midnight Sun.

**aphelion:** the point in an object's orbit where it is farthest from the Sun (helios = sun). The Earth is at aphelion each year on about July 3.

**apparent magnitude:** a system used to compare the apparent brightness of celestial objects. The lower an object's apparent magnitude, the **brighter** it is. A change in magnitude of 1 corresponds to a change in brightness by a factor of 2.5. Objects with a magnitude of less than 6 can be seen with the naked eye in good observing conditions.

**arcminute:** a unit of angular measure equal to one sixtieth of a degree. As an example, the Moon has an apparent diameter of about 30 arcminutes.

**arcsecond:** a unit of angular measure equal to one sixtieth of one sixtieth of a degree, or one sixtieth of an arcminute. As an example, the apparent diameter of Jupiter is about 45 arcseconds.

**Arctic Circle:** the line of latitude on the Earth's surface that is 23-1/2 degrees south of the North Pole. The Arctic Circle marks the southernmost points in the Northern Hemisphere that experience the midnight Sun.

**asterism:** a group of stars that people informally associate with each other to make a simple pattern, such as the Big Dipper and Square of Pegasus. The stars in an asterism can come from one or more official constellations.

**asteroid:** one of the many thousands of chunks of rock or iron that orbit the Sun, also known as minor planets (an older term). Most orbit between Mars and Jupiter, where they formed, but some cross the orbit of the Earth. Fragments of asteroids are called meteorites if they fall to the ground.

**astrological sign:** one of 12 sections of the zodiac that are 30° long and that corresponds to the positions of the constellations as they were about 2,600 years ago when the astrological system was established. Do not confuse an astrological sign with an astronomical constellation with the same or a similar name, as they only partially overlap.

**autumnal equinox:** the moment when the Sun crosses the ecliptic in a southward direction on or about September 22. In the Northern Hemisphere, it marks the first day of autumn. It is also the Sun's position in the sky at that moment. In the sky, it is one of two intersection points of the ecliptic and celestial equator (the other being the vernal equinox).

**binary star:** two stars which orbit a common center of gravity. These stars often look like a single star to the naked eye. If more than two stars orbit a common center of gravity, it is called a multiple star.

**birth sign:** see *astrological sign*.

**celestial equator:** the projection of the Earth's equator into space; also a line in the sky midway between the North and South Celestial Poles. The celestial equator is the line of zero declination in the equatorial coordinate system.

**celestial meridian:** the line of zero right ascension in the equatorial coordinate system.

**celestial sphere:** the projection of the Earth into space. The stars can be imagined to be drawn on the inside of this sphere.

**circumpolar:** those stars that are so far north (or so far south, in the Southern Hemisphere) they do not set at all as seen from a given latitude.

**comet:** a body composed of ice and dust in orbit around the Sun.

**conjunction:** the passing of one planet by another planet or by the Moon or Sun. Two planets are in conjunction when they have the same ecliptic longitude (or alternately the same right ascension).

**constellation:** one of the 88 portions of the sky that are officially recognized by the International Astronomical Union (see Appendix A). A constellation is an arbitrary area of the sky, and it includes everything within that area's boundaries, regardless of distance from the Earth.

**Daylight Saving Time:** the adjustment to the clock time that is put into effect during the summer to extend daylight one hour later in the evening. In the United States and Canada, Daylight Saving Time begins on the first Sunday in April (set clocks ahead one hour) and ends on the last Sunday in October (set clocks back one hour); other countries change on different dates. Daylight time is one hour ahead of the equivalent standard time (6 P.M. standard time becomes 7 P.M. when daylight time is in effect).

**declination:** the angular distance of an object north of (positive) or south of (negative) the celestial equator, expressed in degrees. It is the celestial equivalent of latitude on the Earth's surface. The declination of the celestial equator is 0°; the declination of the North Celestial Pole is +90°, and the declination of the South Celestial Pole is -90°.

**double star:** two stars that appear near each other in the sky. Their apparent closeness may be due to chance alignment, with one star far beyond the other, or they may be in orbit around a common center of gravity, in which case they form a *binary star*.

**eclipse:** the passage of one object in front of another (as the Moon passes in front of the Sun during an eclipse of the Sun), or the passage of one object through the shadow of another (as the Moon passes through the shadow of the Earth during an eclipse of the Moon).

**eclipse season:** the 38-day period when the Sun is near a node of the Moon's orbit and one or more solar eclipses may happen.

**ecliptic:** the Sun's apparent annual path through the fixed stars; also the orbit of the Earth if it could be seen in the sky. The 13 astronomical constellations that the ecliptic passes through are the astronomical constellations of the zodiac (12 in astrology).

**ecliptic coordinate system:** the system of specifying positions in the sky that uses the ecliptic – the Sun's apparent path around the sky – as the fundamental reference plane. Ecliptic coordinates are useful when specifying positions in the solar system and especially positions relative to the Sun.

**ecliptic latitude:** the angular distance of an object above (positive) or below (negative) the ecliptic, expressed in degrees. The ecliptic latitude of the Sun is always zero.

**ecliptic longitude:** the angular distance of an object, measured along the ecliptic, eastward from the vernal equinox, and expressed in degrees. The ecliptic longitude of the Sun is 0° when the Sun is on the vernal equinox, and it increases by very nearly 1° per day through the year.

**elliptical galaxy:** one of the two major types of galaxies, the other being a *spiral galaxy*. They often resemble star clusters when seen from the Earth.

**epoch:** particular date for which astronomical positions in a book or table are accurate. Most books give star positions which are valid for the J2000 (January 1, 2000) epoch.

**equation of time:** the difference between true solar time (determined by the Sun's position in the sky) and mean solar time (the time told by your watch). The two times can vary by as much as 16 minutes over the course of a year.

**equatorial coordinate system:** a system of specifying positions in the sky that uses the celestial equator – the projection of the Earth's equator into space – as the fundamental reference plane.

**field of view:** angular width of sky that can be seen with an optical instrument. Field of view is measured in degrees, arcminutes, and arcseconds.

**Foucault pendulum:** pendulum which varies the direction of its swing as the Earth rotates. Used to demonstrate that the Earth rotates, not the sky.

**full Moon:** the Moon when it lies directly opposite the Sun. The Moon is full two weeks after new Moon. The full Moon rises at sunset and sets at sunrise. The Earth is between the full Moon and the Sun.

**galactic coordinate system:** the system of specifying positions in the sky that uses the plane of the Milky Way as the fundamental reference plane.

**globular cluster:** a huge spherical cluster of tens of thousands of stars. The stars of a cluster were born together and travel through space together. M13 and M22 are familiar examples.

**greatest eastern elongation:** the greatest angular distance to the east of the Sun reached by Mercury or Venus. When a planet is at its eastern elongation, it sets after the Sun and is at its best visibility in the evening sky.

**inferior conjunction:** the passage of Mercury or Venus between the Earth and the Sun. The outer planets cannot pass between the Earth and the Sun and cannot come to inferior conjunction.

**inferior planet:** Mercury or Venus, so-called because their orbits are inside the Earth's orbit around the Sun and "inferior" to the Earth in terms of distance from the Sun.

**Julian Day:** the number of days (and fractions of days) that have elapsed since noon, Jan.1, 4713 BC (Greenwich Mean Time). It is used to simplify calculating the time interval between two events.

For example, 9:00 P.M. P.S.T. on January 1, 2000, was Julian day 2,451,545.71.

**latitude:** the angular distance of an object north or south of the Earth's equator expressed in degrees. The latitude of the equator is 0°, of Chicago is 42° North, the North Pole is 90° North, and Lima, Peru is 12° South. Latitudes south of the equator are expressed as South or negative.

**light year:** the distance light travels in one year. One light year equals 9,460,536,000,000 kilometers, or 5,878,507,000,000 miles.

**limiting magnitude:** the magnitude of the dimmest object that can be seen through an optical instrument (including the eye). The limiting magnitude of an instrument will vary with light conditions.

**light pollution:** the brightening of the night sky due to artificial light. Light pollution makes it impossible to view many dim objects that can only be seen in a dark sky.

**locked rotation:** condition where a moon has the same period of rotation as its period of revolution around its parent body. This means that the Moon always shows the same face to its parent planet. Our Moon is in locked rotation around the Earth.

**longitude:** the angular distance of an object east or west of the Prime Meridian (the line of zero longitude which runs through Greenwich, England), expressed in degrees. The longitude of Chicago is 88° West.

**luminosity:** an expression of the true brightness of a star as compared to the Sun. The Sun's luminosity is 1.0 by definition. Sirius has a luminosity of 23 and Rigel a luminosity of about 50,000.

**lunar eclipse:** an eclipse of the Moon (lunar = moon), caused when the Moon moves partially or wholly into the shadow of the Earth and grows dark for up to a few hours. A lunar eclipse can be seen by everyone on the side of the Earth facing the Moon.

**magnitude:** an expression of the brightness of a star (or other celestial object) as it appears from the Earth, according to a system devised by Hipparchus. Also known as apparent magnitude, to distinguish from absolute magnitude. Larger numbers refer to fainter stars, and the brightest stars and planets have negative magnitudes. One magnitude difference is equal to a brightness difference of 2.5 times.

**massing:** close alignment of three or more planets (or two or more planets and the Moon), as seen from Earth. This occurs when all the bodies involved in the massing have similar ecliptic longitudes.

**meridian:** the line in the sky that extends from the southern point on the horizon through the zenith (overhead point) to the northern point on the horizon, bisecting the sky into an eastern and western half. Objects are at their highest when they cross the meridian; the Sun is on the meridian at local noon. Also a line on the surface of the Earth (or another body) that extends from pole to pole.

**Messier object:** one of the 110 objects in the catalog compiled by Charles Messier. Most Messier objects are galaxies, star clusters, or nebulae.

**meteor:** the visible flash of light produced when a meteorite falls through the atmosphere and bursts into flame because of friction with air molecules; also called a "shooting star" or "falling star."

**meteorite:** the solid particle, either stone or iron, that falls through the atmosphere to produce a meteor. Most meteorites are fragments of asteroids. Science museums display meteorites that survived their falls.

**midnight Sun:** the Sun when visible at midnight, which happens only in summer north of the Arctic Circle or south of the Antarctic Circle.

**Milky Way:** our own galaxy.

**minor planet:** *see asteroid.*

**month:** the period of time it takes the Moon to orbit the Earth (or a moon to orbit a planet). A sidereal month (27-1/3 days) is the time it takes the Moon to orbit the Earth and return to the same position relative to the stars; a synodic month (29-1/2 days) is the time it takes the Moon to orbit the Earth and return to the same position relative to the Sun and is the time between new Moon and new Moon.

**nadir:** the point in the sky directly beneath an observer's feet, opposite of zenith.

**nebula:** a cloud of gas or dust in space, either between the stars or expelled by a star; nebula is Latin for cloud. There are many kinds of nebulas.

**new Moon:** the Moon when it lies in the same direction as the Sun and the beginning of a cycle of lunar phases. The new Moon rises and sets with the Sun. The Moon is between the Earth and the Sun at new Moon.

**night vision:** enhanced ability to see objects in the dark. Night vision is ruined if the eyes are exposed to bright light.

**node:** the point(s) in the sky where two orbits or paths cross. The nodes of the Moon's orbit are the two places where the Moon's orbit crosses the ecliptic.

**North Celestial Pole:** the sky's north pole; the point in the sky directly above the Earth's north pole.

**obliquity of the ecliptic:** the amount of the tilt of the Earth's axis (23.5°) which determines the angle the ecliptic makes with the celestial equator as they intersect in the sky.

**occultation:** the disappearance (eclipse) of one object behind another, as a star or planet behind the Moon or a star behind a planet.

**open cluster:** a diffuse association of a few dozen to a few thousand stars, all of which were born together and which travel through space together. The Pleiades in Taurus and Beehive in Cancer are familiar examples.

**opposition:** the position of a planet when it is opposite the Sun in the sky. Only objects that orbit outside the Earth's orbit come into opposition; Mercury and Venus cannot.

**parallax:** the apparent shift in position of an object when it is viewed from two different points. The parallax of a star is measured from opposite ends of the Earth's orbit, and for the nearest stars is less than one second of arc (one arcsecond).

**parsec:** the distance an object would have if its parallax were one second of arc (see *parallax*). One parsec equals 3.26163 light years or 30,856,780,000,000 kilometers.

**penumbra:** shadowed area in an eclipse where only part of the light source is blocked. Observers in the penumbral shadow of a solar eclipse see a partial eclipse.

**perihelion:** the point in an object's orbit when it is closest to the Sun (helios = sun). The Earth is at perihelion each year on about January 3.

**period:** the time an object takes to complete a certain motion and return to its original state, e.g. period of revolution.

**planetary nebula:** a luminous cloud of gas expelled by an aging star that has become unstable. The name comes from a nebula's superficial resemblance to the faint planets Uranus and Neptune as seen through a small telescope. The Ring Nebula M57 in Lyra is a familiar example.

**precession of the equinoxes:** the slow wobbling of the Earth's axis in a 25,800-year cycle, caused by the gravitational attraction of the Moon on the Earth's equatorial bulge. Precession causes the vernal equinox (and all other points on the ecliptic) to regress westward along the ecliptic, slowly changing the equatorial coordinate grid.

**Prime Meridian:** the line of longitude which passes through Greenwich, England, and which is the zero line for expressing longitude on the Earth's surface.

**proper motion:** the motion of the stars relative to each other, caused by their actual motion in different directions at different speeds through space.

**retrograde motion:** the apparent "backwards" or westward motion of a superior planet against the background of the stars caused when the faster-moving Earth on an inside orbit passes that planet.

**revolution:** the orbiting of one body around another body, as the Moon revolves around the Earth and the Earth revolves around the Sun.

**right ascension:** in the equatorial coordinate system, the angular distance of an object eastward from the zero point (which is the vernal equinox), usually expressed in hours and minutes (which represents the Earth's rotation from the vernal equinox to the object). It is the celestial equivalent of longitude on the Earth's surface.

**rotation:** the spinning of a body on its own axis. The Earth rotates once a day. See *revolution*, which is often confused with rotation.

**sidereal day:** the time it takes the Earth to rotate once relative to the stars, in 23 hours 56 minutes and 4 seconds, which is 4 minutes less than a solar day. (During one sidereal day the Sun moves 1° east along the ecliptic, and the Earth has to rotate 4 additional minutes to complete one rotation relative to the Sun in one 24-hour solar day.)

**sidereal month:** the time it takes the Moon to complete one orbit of the Earth and return to the same position among the stars, on an average of 27.32166 days.

**sidereal year:** the time it takes the Earth to complete one orbit of the Sun relative to the stars, in 365 days 6 hours 9 minutes and 10 seconds. It is also the time it takes the Sun to appear to travel once around the sky relative to the stars.

**solar day:** the time it takes the Earth to spin once relative to the Sun, in exactly 24 hours (by definition).

**solar eclipse:** an eclipse of the Sun by the Moon, when the Moon passes in front of the Sun. Solar eclipses can be partial, total, or annular. Only the few people in the narrow "path of totality" see a solar eclipse as total.

**South Celestial Pole:** the point in the sky directly above the Earth's south pole.

**spiral galaxy:** The second major type of galaxy, characterized by a central bulge and a number of spiral arms extending from the bulge.

**standard time zone:** one of the 24 sections of the Earth, each about 15 degrees wide and extending from pole to pole, within which the time is the same. In practice, natural and political boundaries determine the edges of time zones.

**summer solstice:** the moment when the Sun reaches its greatest distance north of the celestial equator, on or about June 21. In the Northern Hemisphere this marks the first day of summer; in the Southern Hemisphere it marks the first day of winter.

**superior conjunction:** the position of a planet when it is on the far side of the Sun (and in conjunction with the Sun).

**superior planet:** the planets Mars through Pluto, so-called because their orbits are outside the Earth's orbit around the Sun and thus "superior" to the Earth in terms of distance from the Sun.

**synodic month:** the time it takes the Moon to complete one cycle of phases, such as from new Moon to new Moon, on an average of 29.53059 days.

**transit:** the passage of one celestial body across the face of a second larger body.

**triple conjunction:** close alignment of a planet and a star at three distinct times, caused by the retrograde motion of the planet. The planet passes the star once in its forward motion, once more in its retrograde motion, and a third time when it resumes its forward motion.

**tropical year:** the length of time it takes the Sun to circle the sky relative to the vernal equinox. The tropical year is identical to our standard "year."

**Tropic of Cancer:** the line of latitude on the Earth's surface that is 23-1/2 degrees north of the equator. It marks the northernmost points in the Northern Hemisphere from which the Sun can appear directly overhead.

**Tropic of Capricorn:** the line of latitude on the Earth's surface that is 23-1/2 degrees south of the equator. It marks the southernmost points in the Southern Hemisphere from which the Sun can appear directly overhead.

**umbra:** shadowed area in an eclipse where the light source is completely blocked. Observers in the umbral shadow of a solar eclipse experience a total eclipse.

**Universal Time:** in simplest terms, the time at the longitude of Greenwich, England. Universal Time (UT) is widely used in international publications as a standard time.

**variable star:** a star whose light changes over time. Stars can vary in brightness for a variety of reasons, from eclipses by companions to instability of their interiors that causes the stars to swell and shrink.

**vernal equinox:** the moment when the Sun crosses the ecliptic in a northward direction on or about March 21. In the Northern Hemisphere, it marks the first day of spring. It is also the Sun's position in the sky at that moment. In the sky, it is one of two intersection points of the ecliptic and celestial equator (the other is the autumnal equinox).

**waning crescent:** the phase of the Moon between third quarter and new Moon. Waning means declining or fading.

**waning gibbous:** the phase of the Moon between full Moon and last quarter.

**waxing crescent:** the phase of the Moon between new Moon and first quarter. Waxing means increasing.

**waxing gibbous:** the phase of the Moon between first quarter and full.

**winter solstice:** the moment when the Sun reaches its greatest distance south of the celestial equator, on or about December 22. In the Northern Hemisphere this marks the first day of winter; in the Southern Hemisphere it marks the first day of summer.

**zenith:** the point in the sky directly above the observer; the top of the sky.

**zodiac:** the band of constellations or signs that the Sun passes through as it moves around the sky. There are 12 signs of the astrological zodiac but 13 constellations of the astronomical zodiac.

# INDEX

Page numbers in *italics* indicate illustrations.
For definitions see glossary.